Building a Release Pipeline with
Team Foundation Server 2012

Building a Release Pipeline with Team Foundation Server 2012

Larry Brader
Roberta Leibovitz
Jose Luis Soria Teruel

"This book provides an excellent guide to the principles and practices of continuous delivery. If you're using Microsoft's toolchain, you'll find this book an indispensable resource."

Jez Humble
Principal, ThoughtWorks

ISBN: 978-1-62114-032-0

Contents

Foreword

The software industry has gone through a rapid evolution over the past few years with product requirements becoming increasingly sophisticated and customer expectation around delivery time getting shorter. Shipping software faster is an aspiration that all of us in software development teams universally harbor. The ability to provide uninterrupted flow of customer value from inception to production sounds like magic. Doesn't it? Well, it no longer has to be. Over the years, there have been several examples of teams reaching the near Zen-like state of continuous delivery using some simple and fundamental principles of releasing software.

This book covers those principles in detail and more importantly shows the application of those principles in a real world scenario. It walks through the journey of a software development team striving to realize a goal that is universal in nature—ship products on time, within budget. If you have ever been part of a software development team trying to get better at shipping software, I am sure you will find yourself nodding your head at the situations they encounter along the way.

Cultural change is the most important and challenging part of bringing your teams together to deliver quality software faster. It may sound cliché, but the biggest enemy here is the siloes we built in our teams in the hope of optimizing for efficiency. Building software is a team sport. We need to acknowledge that and act in ways that reinforce the same. Bringing together development, QA and operations teams on a shared goal but seemingly contrary requirements is a critical part of making this change successful.

Once the team is set up to operate in a friction-free manner, tooling is the next most important piece of the puzzle. Having all team members speak a common language, focus on a common set of metrics, and plugged into a common system that helps visualize progress on the shared goal is key.

Visual Studio 2012 provides developers a powerful toolset to set up a simple, integrated continuous delivery pipeline to manage software releases. Starting from problem definition and visualization to orchestrating the release through various stages, automating the whole process for efficiency and finally releasing high quality software, the Visual Studio product line has tooling to accomplish each of these steps easily and efficiently.

Like all things in the high-tech industry, Visual Studio has undergone rapid and significant changes in the release management toolset available with it. The book uses Visual Studio 2012 as the toolset illustrated, but the newly released Visual Studio 2013 works equally well, in fact better, for the examples given in the book. I strongly recommend you do the labs and exercises in the book as you read each chapter to fully appreciate the essence of the exercises

As you trace through the struggles of the Trey Research team and how they overcome challenges at each stage to successfully deliver on their project, do reflect on similar situations on your team and explore ways to apply the insight you received from reading this book. If it takes you one step closer in reality on your path to shipping great software faster, the authors would have successfully accomplished what they set out to do.

Anutthara Bharadwaj
Principal Group Program Manager, Visual Studio ALM

The Team Who Brought You This Guide

This guide was produced by the following individuals:

- Program and Product Management: Larry Brader (Microsoft Corporation)
- Subject matter expert writer: Jose Luis Soria Teruel (Plain Concepts Corporation);
- Writer: Roberta Leibovitz (Modeled Computation, LLC)
- Development and test: Larry Brader and Kirpa Singh (Microsoft Corporation); Poornimma Kaliappan (VanceInfo)
- Edit: RoAnn Corbisier (Microsoft Corporation)
- Cartoons: Paul Carew (Linda Werner & Associates Inc)
- Book layout and technical illustrations: Chris Burns (Linda Werner & Associates Inc.)
- Release management: Nelly Delgado (Microsoft Corporation)

We want to thank the customers, partners, and community members who have patiently reviewed our early content and drafts. Among those, we want to highlight the exceptional contributions from our Advisor Council and the ALM Rangers.

Advisor Council

Tiago Pascoal, Ian Green, Paul Glavich, Matteo Emili, Perez Jones Tsisah, Marce o Hideaki Azuma, Arnoud Lems, Paulo Morgado, Bruce Cutler, Mitchel Sellers, and Aleksey Sinyagin

ALM Rangers

Big thanks to _Willy-Peter Schaub_ for his input and coordinating the ALM Rangers in contributing to the book.

ALM Ranger Subject Matter Experts

Casey O'Mara , _Jeff Bramwell_, _Krithika Sambamoorthy_, _Michael Fourie_ and _Micheal Learned_

ALM Ranger Reviewers

Andrew Clear, _Anna Galaeva_, _David Pitcher_, _Francisco Xavier Fagas Albarracin_, _Gordon Beeming_, _Hamid Shahid_, _Hassan Fadili_, _John Spinella_, _Mathias Olausson_, Mehmet Aras, _Richard Fennell_, Tiago Pascoal, Tommy Sundling, and Vlatko Ivanovski

1   You Want It When?

Does this sound familiar? You're expected to produce releases at an ever-increasing rate. You're under pressure to add new features and deploy to customers sometime between your first cup of coffee in the morning and lunch, if you have time to eat it. In the meantime, you have the same release processes you've always had and it's got problems. Maybe there's some automation, but there's room for lots of improvement. Manual steps are everywhere, everyone has a different environment, and working all weekend to get a release into production is normal.

One of the biggest problems is that changing how your software is released won't happen by waving a magic wand or writing a memo. It comes through effort, time, and money. That takes commitment from every group involved in the software process: test, development, IT (operations), and management. Finally, change is scary. Your current release process bears no similarity to the well-oiled machines you've seen in a dozen PowerPoint presentations, but it's yours, you know its quirks, and you are shipping.

This guidance is here to help you with some of these challenges. It explains how to progressively evolve the process you use to release software. There are many ways to improve the release process. We largely focus on how to improve its implementation, the release pipeline, by using and customizing the default build templates provided by Team Foundation Server (TFS) and Lab Management. We move forward in small iterations so that no single change you make is too drastic or disruptive.

The guidance also shows you how to improve your release process by using some of the tools that TFS offers. For example, it shows you keep track of your product backlog and how to use Kanban boards.

The goal of this guidance is to put you on the road toward continuous delivery. By continuous delivery, we mean that through techniques such as versioning, continuous integration, automation, and environment management, you will be able to decrease the time between when you first have an idea and when that idea is realized as software that's in production. Any software that has successfully gone through your release process will be software that is production ready, and you can give it to customers whenever your business demands dictate. We also hope to show that there are practical business reasons that justify every improvement you want to make. A better release process makes economic sense.

THE RELEASE PIPELINE

In the abstract, a release pipeline is a process that dictates how you deliver software to your end users. In practice, a release pipeline is an implementation of that pattern. The pipeline begins with code that's in version control (we hope) and ends with code that's deployed to the production environment. In between, a lot can happen. Code is compiled, environments are configured, many types of tests run, and finally, the code is considered "done." By done, we mean that the code is in production. Anything you successfully put through the release pipeline should be something you'd give to your customers. Here's a diagram based on the one you'll see on Jez Humble's *Continuous Delivery* website. It's an example of what can occur as code moves through a release pipeline.

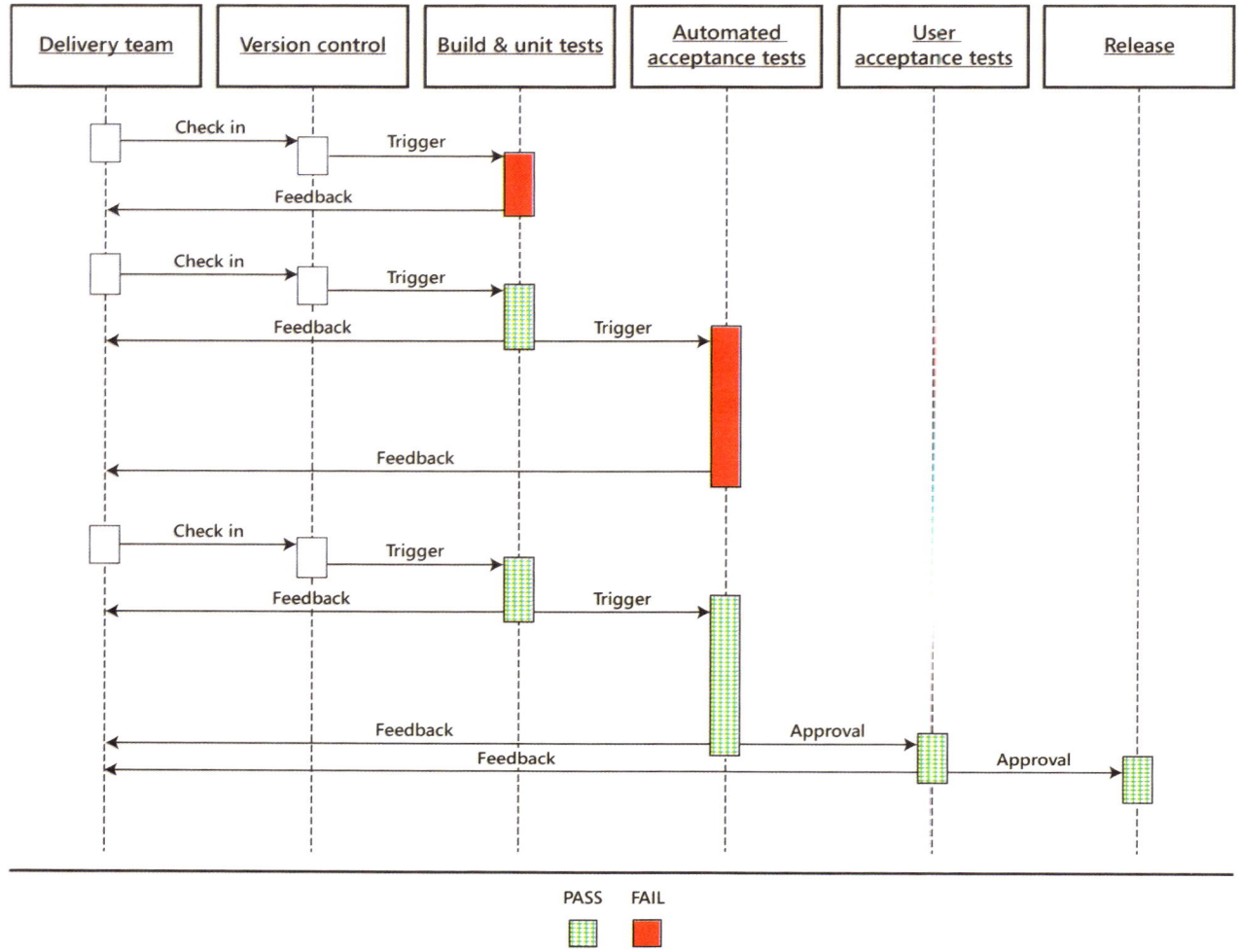

(You should, of course, tailor this pipeline to your own situation, perhaps by adding a variety of other tests.) Notice that every check-in to version control sets the pipeline in motion. If at any point in the pipeline there's a failure, the build goes no further. In general, people shouldn't check in anything else so long as the build and unit tests fail. Some people enforce this by rejecting commits from anyone but the person fixing the build.

The goal is to release your software as soon as possible. There are practices you can follow that will help you do this.

Version Everything

Version all the information that affects your production system. Use a version control system for your source code, certainly, but it can also contain your tests, your scripts, your configuration files, and anything else you can think of that affects your project. You may want to use virtualization libraries such as System Center Virtual Machine Manager (SCVMM) or Windows Azure management tools for virtual environments. For physical environments or imaging and deployment tools for physical environments you might want to consider the _Windows Automated Installation Kit_ (Windows AIK). NuGet might be a good choice as an artifact repository for binaries and dependencies. For more information, go to _http://www.nuget.org/_. SharePoint is used by many teams for their documentation. In fact, any versioning tool you're comfortable with is fine as long as it supports a release pipeline with some automation and is well understood by your team. For more information, go to the _SharePoint_ product site.

Use Continuous Integration

Continuous integration is defined in various ways by various groups. In this book, we use the definition given by Martin Fowler: *Continuous Integration is a software development practice where members of a team integrate their work frequently, usually each person integrates at least daily–leading to multiple integrations per day. Each integration is verified by an automated build (including test) to detect integration errors as quickly as possible. Many teams find that this approach leads to significantly reduced integration problems and allows a team to develop cohesive software more rapidly.*

In this guidance, we mean that you should frequently integrate your work with the main branch. Ideally, explicit integration phases are, at some point, no longer necessary because your code is always integrated.

Use Automation

Wherever you can, automate the release pipeline. Automation makes the release process a repeatable, predictable experience. Think about automating not just the pipeline itself, but how you do provisioning, how you create environments, and even how you maintain your infrastructure. Manual steps are repetitious and error prone while automation makes a process repeatable and reliable.

There are sound business reasons for using automation. It maximizes the talents of the people you've hired and frees them to do what they do best—tasks that are creative and innovative. Leave the drudgery to your computers. They never get bored. Automation helps to remove dependencies you might have on particular people, who are the only ones who can, perhaps, deploy to the production environment, or run some group of tests. With automation, anyone with the correct permissions can set the process in motion.

Manage Environments

Are your developers and testers handcrafting their own environments, manually installing each piece of software and tweaking configuration files? How long does it take for them to do this? Managing your environments by using automation can solve many problems that plague teams as they try to release their software.

Automation can help to create environments that conform to some known baseline. Automation also makes your environments as versionable, repeatable, and testable as any other piece of software. Finally, it's much easier to create environments with automation, which, in turn means that by making environments (and the tools that create them) available early, every team member can run and test the code in consistent and stable environments from the onset of the project.

If you can, keep the differences between each environment as small as possible. The closer the environments are to each other, the easier it will be to achieve continuous delivery because you can identify interoperability conflicts between the code and the environment long before you reach production. If you do have differing environments (this can be particularly true for development environments), have your key testing environments mirror the production environment as closely as possible.

For some people, the amount of time it takes for a developer to set up a machine is a litmus test that indicates how difficult it's going to be to start automating other aspects of the release pipeline. For example, if a developer can set up a system in a few hours or less, then there's probably some processes and tools already in place that will help with the rest of the automation. If it takes more than a day then this could indicate that automation is going to be difficult.

Fail Fast, Fail Often

Failure shouldn't be feared. You can't innovate or learn without it. Expect it, and resolve the issues when they arise.

To address problems quickly, you need to know that a problem is there as soon as possible. Every validation stage should send feedback to the team immediately if the software fails. Additionally, the tests themselves should run quickly. This is particularly true for the unit tests. These initial tests should complete in a few minutes. If your software passes, you have a reasonable level of confidence that it works. If it fails, you know the software has a critical problem.

The other test stages may run slowly. If they take a very long time, you might want to run them in parallel, across multiple machines rather than on a single machine. Another possibility is to make the pipeline wider rather than longer, breaking the dependencies that are inherent in a strictly sequential system. Here's an example that shows a sequential pipeline.

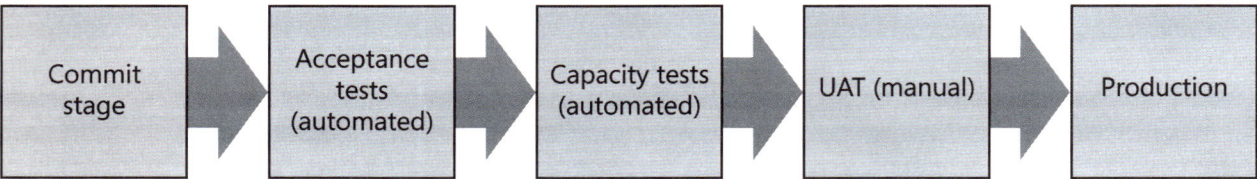

In this pipeline, one stage follows another. If the acceptance tests, for example, take a long time to run, then capacity testing is delayed until they finish. You may be able to rearrange some of the stages so that all builds that pass some designated stage are available. Here's an example that shows the same pipeline, but now shorter and wider.

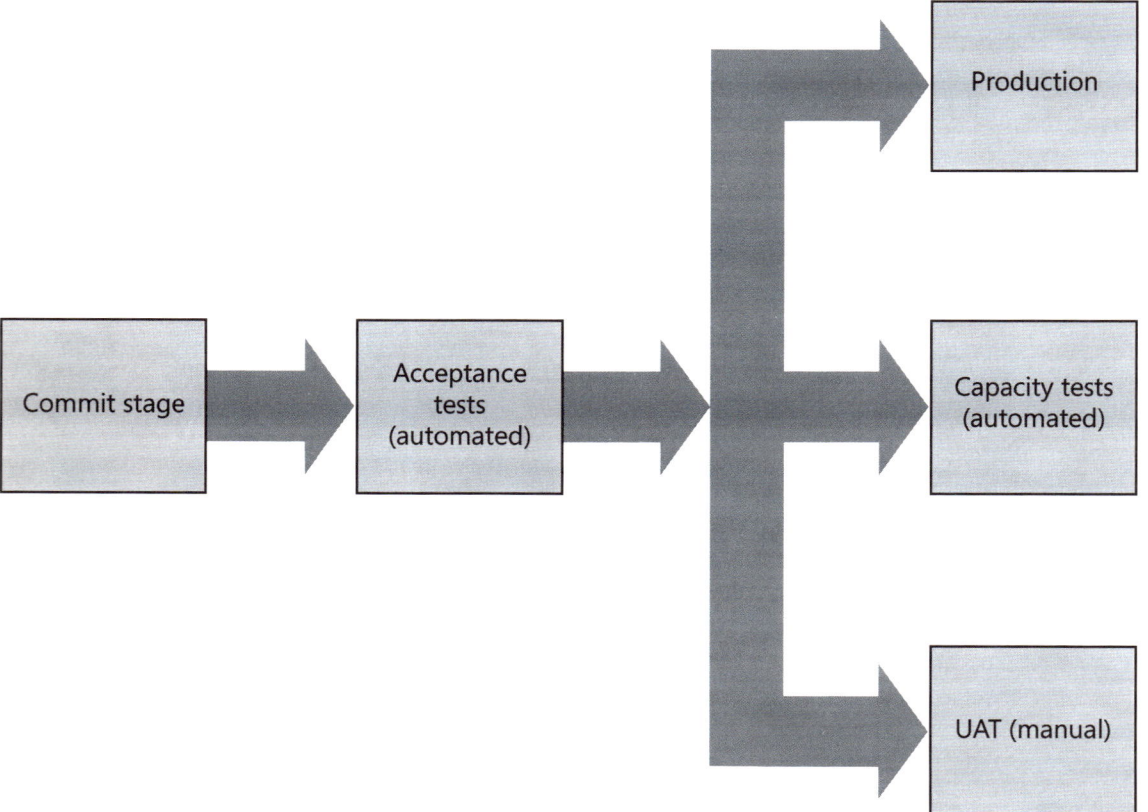

Any build that passes the acceptance tests can go to production, undergo automated capacity tests or be proven to meet all contractual requirements with manual user acceptance tests (UAT). Breaking dependencies by stacking your pipeline gives you more flexibility than a sequential pipeline does. You can react more quickly to circumstances, such as the need to quickly release a hotfix or bypass an unnecessary stage.

Provide Visibility

Visibility means that everyone on the team has access to all the information required to know what's happening to the software as it goes through the pipeline. Examples of what you might want to know include the build's version number, the build configuration, and the tests that failed. How you expose the information is up to you. You may have a dashboard, you may use a whiteboard, but whatever method you choose, all team members should have easy access to the information.

Some people refer to the display that makes the information visible as an _information radiator_, a term first coined by Alistair Cockburn. According to Cockburn, "an information radiator is a display posted in a place where people can see it as they work or walk by. It shows readers information they care about without having to ask anyone a question. This means more communication with fewer interruptions." Qualities of a good radiator are:

- It's large and easily visible to the casual, interested observer.
- It's understood at a glance.
- It changes periodically, so it's worth visiting and revisiting.
- It's easily kept current.

People get very creative when they design their radiators. They use computer screens, wall boards with sticky notes and even lava lamps. One popular approach is to use a traffic light, with four possible combinations.

If the light is green then the build and all the tests have passed. If the light is yellow, then the build and tests are in progress. If both the yellow and green lights are on, then the build is unlikely to fail. If the light is red, some part of the build or the tests has failed.

Bring the Pain Forward

If there's something particularly painful in your release process, do it more frequently and do it sooner. Front load your pipeline so the hardest steps happen early. For example, if you do most of your testing at the end of the project and this isn't working well for you, consider doing many of the tests early, as soon as a commit happens.

If you've begun to increase the number of releases or the pace at which you're creating releasable software, you may find that quality assurance (QA) and information security (Infosec) groups are lagging behind. Perhaps it takes several months for Infosec to perform a review. If this is the case, the answer is still the same. Start incorporating security tests into the integration process instead of waiting until the end of the project. If static code analysis tools are taking too long, perform the analysis on every check-in for only the most important set of rules. Run the rest of the validations as early and as often as possible. You may even want to have a dedicated code analysis stage that performs exhaustive tests. Static code analysis is performed on the assemblies, so you won't have to build again to perform the tests. Again, perform the less critical analyses as early and as often as possible.

Take Small Steps

Even one of guidelines we've discussed might sound difficult to implement, let alone all of them. Try to identify a single aspect of your release process that you'd like to improve. Perhaps take a look at the one that's giving you the biggest problems. Talk it over with your team and think about a feasible solution that would improve matters even a little. Implement it. Did it work? Is life better? If not, why not? If it did work, do the same thing again for another problem.

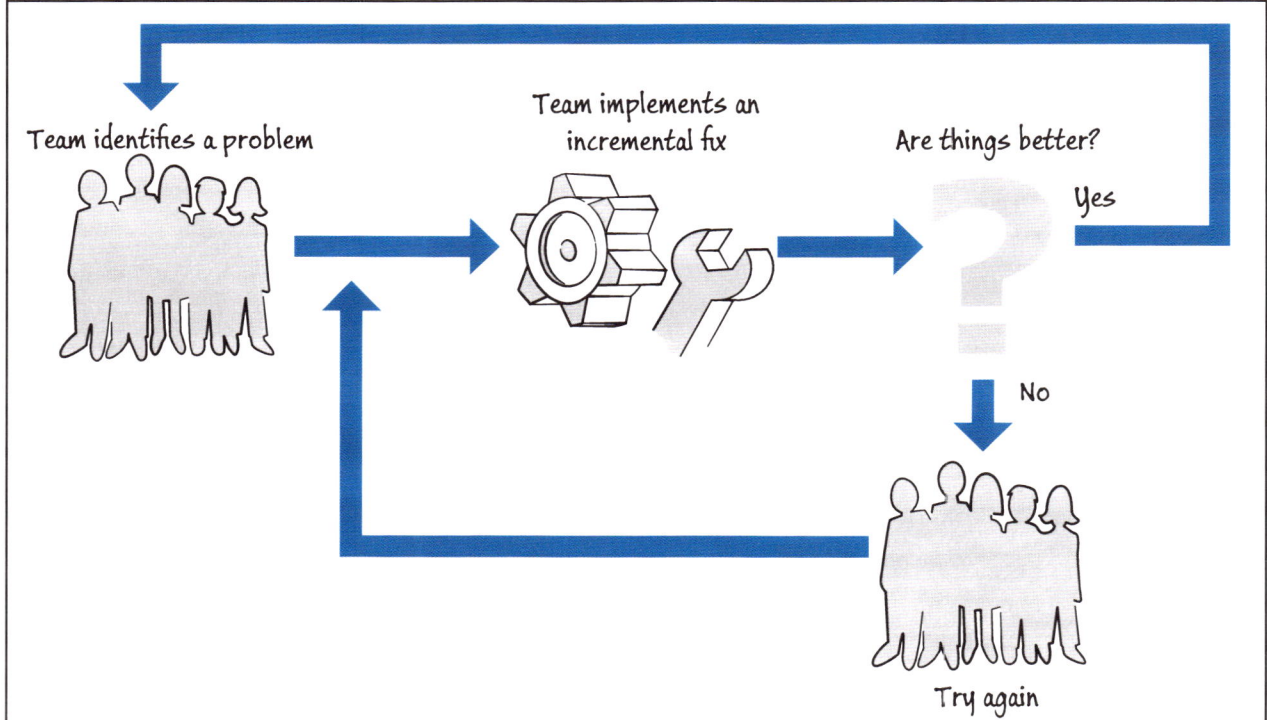

This cycle of iterative process management is often called the Deming cycle or the PDCA (plan-do-check-adjust) cycle. Edward Deming is considered by many to have initiated this modern quality control movement. The article in Wikipedia on *PDCA* gives an introduction to the subject.

Think About DevOps

The goals and practices we've discussed are often spoken of in terms of a software development mindset called DevOps. DevOps, an outgrowth of the Agile movement, stresses cooperation and communication between everyone involved in the development and release of good software. The name itself is a combination of development and operations (IT professionals), probably because these two groups often find themselves at odds with each other. Developers are rewarded according to how many new features they can create and release. Ops people are rewarded according to how stable and secure they can make the company's infrastructure. Developers may feel that Ops is slow and stodgy. Ops may feel that developers don't appreciate what it takes to actually release new software, let alone maintain what's already there.

However, it isn't only operations teams and software developers who are involved in the process. Testers, database managers, product and program managers, anyone involved in your project, should be a part of the release process. DevOps stresses close collaboration between traditionally distinct disciplines or silos.

This book touches on some of the principles espoused by DevOps proponents. It uses a fictional company, Trey Research, as the setting and, as you'll see, the employees of Trey Research find that building good software is about more than the tools. There's a very human component as well.

IS IT WORTH IT?

Improving the release pipeline isn't easy and a good question that you, or your managers might ask is "Is it worth it?" The most direct answer is in the Agile manifesto, published in February, 2001. Its first principle is "Our highest priority is to satisfy the customer through early and continuous delivery of valuable software." This statement is the justification for improving your release process. To point out the obvious, businesses thrive when their customers are happy. If they're not, they'll start looking elsewhere for answers. In support of that goal, improving your release process can result in:

- Faster time to market
- Better quality software
- More productive employees

Faster Time to Market

Leaders in the world of online businesses have shrunk the timeline for software delivery from months to days or even hours. No matter what size business you have, customers now expect features such as real-time customer service and frequent releases of services. In his talk "*Velocity Culture*" given at Velocity 2011, Jon Jenkins, at that time a director at Amazon.com, announced that Amazon was deploying every 11.7 seconds. You may not need to be this fast, but if your organization is only releasing twice a year while a competitor is releasing once a month, there's a problem.

Better Quality Software

The more your pipeline can produce predictable, repeatable results, the better your software. Any aspect of improving your pipeline impacts the quality of your software. If you make incremental changes you'll be able to find bugs easier. If you can deploy those changes early, you'll know right away if you're working on the right features. Find out if your customers like what you're doing before you've invested enormous amounts of time and money.

More Productive Employees

If you can reduce the number of repetitive, frustrating tasks your employees have to do, they'll have more time to exercise the talents that were the reasons you originally hired them. If your devs aren't overwhelmed trying to fix bugs from changes they made a month ago, they'll have more time to implement better products and services. If your testers aren't tied up with tests that could be done faster and better by a computer, they'd have time to come up with really creative ways to give the new app a workout. The same is true for everyone involved in releasing the software. People get to do what they're really good at and computers get to deal with all the drudgery.

THE TOOLS YOU NEED

The patterns we discuss in this book hold true everywhere, no matter how you implement them. We also present a particular solution that requires a specific set of tools. Here's what we use to create our pipeline.

Visual Studio 2012 Virtual Machine

The Visual Studio 2012 Application Lifecycle Management Virtual Machine (VM) is the environment you use for all the HOLs that accompany this guidance, except for those labs marked as advanced. This VM is familiarly known as the Brian Keller VM, and that's how we'll refer to it. For a complete description of the VM and instructions on how to download it, see *Brian Keller's blog*.

> All the *hands-on labs* (HOL) that accompany this guidance run on the Visual Studio 2012 VM except for the Windows Phone 8 labs. These labs are considered to be optional and advanced, and they are not supported by the VM. They require Windows 8, and for you to set up a Windows Communication Foundation service on Windows Azure. For more information, see the Introduction document that accompanies the HOLs.

Visual Studio 2012

You're probably already familiar with Microsoft Visual Studio and its integrated development environment (IDE). Visual Studio comes with many tools that can help with, for example, code analysis, testing, and application lifecycle management (ALM). If you want to implement the pipeline we show in this book, you'll need Visual Studio 2012 Ultimate or Visual Studio Premium because we use Visual Studio Lab Management templates and coded UI tests.

Microsoft Visual Studio Team Foundation Server 2012

TFS provides software development teams with the ability to collaborate on their projects. Anyone who is involved with creating software will find tools and capabilities that will help them perform their jobs. By anyone, we mean not just programmers, but testers, architects, program managers, business managers, and others who contribute to the development and release of software. This book stresses the following capabilities:

- Version control. TFS provides a place to store and version source code as well as any other artifacts that impact your software project. Examples of these artifacts include scripts, configuration files, and documentation.
- Test case management. Microsoft Test Management (MTM) stores all the testing artifacts it uses, such as test plans, test cases, bugs, and the results of tests runs in TFS.
- Build automation. TFS lets you automate your builds, which means you assemble your application into a product without human intervention. An automated build can include many activities such as compiling source code, packaging binaries, and running tests. In this guidance we use the TFS build automation system as the basis for the release pipeline's orchestration, stages and steps.
- Reporting. TFS provides many types of reports and metrics that give you insight into all aspects of your project. In this book we concentrate on metrics that help you validate the success of your release pipeline.

- Environment management. TFS, in conjunction with Lab Management, helps you manage and provision your environments. In this book we concentrate on using Lab Management's standard environments as a way of providing consistent environments for everyone involved in the software project.

    Note: *The HOLs that deal with monitoring and metrics have procedures that use TFS reports. TFS reports are only available if you use the full version of Team Foundation Server 2012 and it is installed on Windows Server 2008 or later. To duplicate those procedures and create the reports, you have two options. One is to install the full version of TFS on Windows Server 2008 or later. The other is to use the Brian Keller VM, which already runs on Windows Server.*

Microsoft Test Manager

Microsoft Test Manager (MTM) is the dedicated interface for testers who work with Team Foundation Server. With it, you can create test plans, add and update test cases, and perform manual and automated tests.

Visual Studio Lab Management

Visual Studio Lab Management works with TFS and allows you to orchestrate physical and virtual test labs, provision environments, and automate build-deploy-test workflows. In this book, we use a new feature of Lab Management—standard environments. Standard environments, as opposed to System Center Virtual Machine Manager (SCVMM) environments, allow you to use any machine, whether physical or virtual, as an environment in Visual Studio, Team Foundation Server, and Microsoft Test Manager. Creating standard environments from your current environments is an easy way to get started with Lab Management. You only need to set up a test controller. For a quick tutorial on creating a standard environment, see *Creating a Standard Environment*.

Community TFS Build Extensions

The *Community TFS Build Extensions* are on CodePlex. You can find workflow activities, build process templates, and tools for Team Foundation Build. The pipeline implementation in this guidance uses several of the workflow activities, such as **TFSVersion** and **QueueBuild**.

Web Deploy

Web Deploy is the standard packaging and deployment tool for IIS servers. It includes MS Deploy, which is also used in the HOLs. For more information about Web Deploy, go to the *IIS website*.

Windows Installer XML

The Windows Installer XML (WiX) toolset builds Windows installation packages from XML source code. For more information, go to the *WiX website*.

Microsoft Excel

Portions of the HOLs include data in Excel spreadsheets.

Additional Tools

Two tools have recently become available that are designed to help you deploy a single build to multiple environments.

DevOps Deployment Workbench Express Edition

The ALM Rangers DevOps Deployment Workbench Express Edition is a new tool that can help you to build once and deploy to multiple environments. For more information, see the *ALM Rangers DevOps Tooling and Guidance website*. You can also read Appendix 1 in this guidance to get an overview of what the tool does.

InRelease

InRelease is a continuous delivery solution that automates the release process from TFS to your production environment. By using predefined release paths, InRelease automatically deploys your application to multiple environments. Based on a business-approval workflow, InRelease improves coordination and communication between development, operations and quality assurance to make release cycles repeatable, visible, and more efficient. It gives you a single view of the release process that can help you to identify failures and bottlenecks between stages. Another capability is the ability to perform rollbacks. For more information see the _InRelease website_.

TREY RESEARCH'S BIG DAY

Trey Research is a small startup that makes mobile apps for ecological field work. Its competitors are larger, well-established companies who sell dedicated hardware. Trey Research hopes to succeed by keeping the costs of its products down and by being nimbler than its competitors. Because it produces software, the company wants to be able to quickly add new features in response to customer feedback and shifts in the market.

Trey Research's newest product sends GPS coordinates back to a Windows Communication Foundation service on a Windows Azure Virtual Machine and displays a Bing map on a Windows Phone 8. The app uses Windows Presentation Foundation for its user interface. Today there's a meeting to discuss how the CEO's first demo of the product went at an important conference. Here are the meeting attendees.

Zachary is the CEO of Trey Research. He started as a developer, but found out he was more interested in the big picture. He likes thinking about what software should look like a few years down the road and how his company can be ahead of the pack.

Paulus is a developer who's been working with computers since he was a kid. He has a real passion for code. His hobby is working on open source projects with other programmers from all over the world.

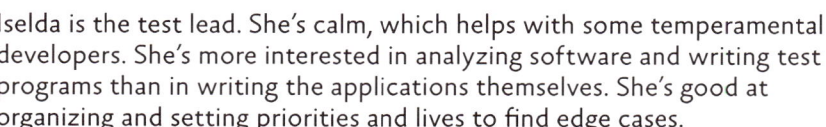

Iselda is the test lead. She's calm, which helps with some temperamental developers. She's more interested in analyzing software and writing test programs than in writing the applications themselves. She's good at organizing and setting priorities and lives to find edge cases.

Raymond is in operations. He likes practical solutions and he's very cautious (although some people might use the word "paranoid"), which makes sense because he's the person who gets the 03:00 call when something goes wrong.

Jin is the new guy. In Chapter 2, he joins Trey Research as a developer. He's worked on all sorts of systems. He likes the idea of being in a small startup where there's lots of opportunity for innovation. He's also a big advocate of continuous delivery and DevOps. He keeps a journal, just for himself, where he records his impressions about what's happening on the new job.

Right now, Raymond, Iselda and Paulus are waiting for Zachary to show up.

Things are not going well for the folks at Trey Research. They have multiple problems, no clear idea why those problems exist, and they're not looking at their situation as a team. The rest of this book is about solving those problems by adopting some new tools and some new ways of working together.

WHAT'S NEXT?

Here's what the rest of this book covers.

Chapter 2: The Beginning

To solve a problem, you first need to analyze what's going wrong. This chapter explains how to develop a *value stream map*, a flow diagram that shows all the steps required to take a product or service from its initial state to the customer. The map includes all the people, processes, times, information, and materials that are included in the end-to-end process. They also start using TFS to manage their projects. They begin to use tools such as a product backlog and a Kanban board.

The chapter's main focus is on the Trey Research release pipeline, as it currently exists. The chapter explains what each stage of the pipeline does, the environments, how code and artifacts are stored, and the tools the Trey Research team uses. Finally, you learn about some of the problems that exist because of how the pipeline is implemented.

Chapter 3: Orchestrating the Release Pipeline

This chapter shows the first steps to take to improve the release pipeline, with continuous delivery as the final goal. It focuses on orchestration, which is the arrangement, coordination and management of the pipeline. You orchestrate the pipeline as a whole and you also orchestrate each stage of the pipeline. A number of best practices are included for guidance. Next, the chapter focuses on the Trey Research team. They decide how to prioritize all the problems they have, and begin to implement changes to their pipeline to address those issues. They use the TFS and Lab Management default build templates to create a skeleton framework that will be the basis for future improvements. They also start to learn about some of the tools TFS offers to manage projects.

Chapter 4: Automating the Release Pipeline

To really make progress, the Trey Research team needs to move away from the largely manual pipeline they have now to one that's largely automated. In this chapter, they automate their deployments, the creation of environments, and at least some of their tests. At the conclusion of this chapter, the team has a fully functional continuous delivery pipeline.

Chapter 5: Getting Good Feedback

The team is celebrating because they now have a fully functional continuous delivery pipeline. They know that their release process is improved, but the problem is that they don't have any actual data that proves it. In this chapter, the team starts to monitor their pipeline so that they can collect all the data it generates and present it in a meaningful way. They also start to track some metrics that are particularly relevant to a continuous delivery release process.

Chapter 6: Improving the Pipeline

The team has gotten a taste for continually improving their pipeline and processes. They know that there is always some area that needs attention. In this chapter, they look at some problems they still have, and consider ways that they can be solved. This chapter deals with Trey Research's future, and what the team can do, over multiple iterations, to make it better.

CONVENTIONS

The guidance contains diagrams of the Trey Research pipeline that show how it changes from iteration to iteration. In the diagrams, we use the color blue to highlight changes in the pipeline. We use a gray italic font to highlight the tools that are used. Here's an example.

COMMIT STAGE

Customized TFS default template

Merge from Dev branch

Get dependencies with NuGet package restore

Perform continuous integration by building the software and running commit tests

Perform code analysis

Perform basic functional tests manually

Version artifacts

Name the pipeline instance (set the build number)

- The commit stage is outlined in blue and its name is in blue because the stage is new.
- The text "Customized TFS default template" is in gray, bold italics because this is a tool that's used for this stage.
- The text "Merge from Dev branch" is in blue because this is a new step.
- The text "Perform code analysis" is in black because it's the same as in the previous iteration.

MORE INFORMATION

There are a number of resources listed in text throughout the book. These resources will provide additional background, bring you up to speed on various technologies, and so forth. For your convenience, there is a bibliography online that contains all the links so that these resources are just a click away. You can find the bibliography at: *http://msdn.microsoft.com/library/dn449954.aspx*.

The book that brought continuous delivery to everyone's attention is Continuous Delivery by Jez Humble and David Farley. For more information, see Jez Humble's blog at *http://continuousdelivery.com/*.

Martin Fowler is another well-known advocate of continuous delivery. His blog is at *http://martinfowler.com/*.

Alistair Cockburn's blog is at *http://alistair.cockburn.us/*.

For guidance that helps you assess where your organization stands in terms of application lifecycle management (ALM) best practices, see the ALM Rangers ALM Assessment Guide at *http://vsaralmassessment.codeplex.com/*.

The ALM Rangers DevOps Deployment Workbench Express Edition can help you to build once and deploy to multiple environments. For more information, see the ALM Rangers DevOps Tooling and Guidance website at *http://vsardevops.codeplex.com/*.

For a complete list of guidance that's available from the ALM Rangers, see the Visual Studio ALM Ranger Solutions Catalogue at *http://aka.ms/vsarsolutions*.

If you're interested in the Edward Deming and the Deming cycle, the article in Wikipedia at *http://en.wikipedia.org/wiki/PDCA* gives an introduction to the subject.

Jon Jenkins's talk "Velocity Culture" is at *http://www.youtube.com/watch?v=dxk8b9rSKOo*.

For more information about the Windows Automated Installation Kit go to *http://www.microsoft.com/en-us/download/details.aspx?id=5753*.

For more information about NuGet, go to *http://www.nuget.org/*.

For more information about SharePoint, go to *http://office.microsoft.com/en-us/microsoft-sharepoint-collaboration-software-FX103479517.aspx*.

The Community TFS Build Extensions are at *http://tfsbuildextensions.codeplex.com/*.

For more information about Web Deploy, go to the IIS website at *http://www.iis.net/downloads/microsoft/web-deploy*.

For more information about WiX, go to the website at *http://wixtoolset.org/*.

To learn about Lab Management standard environments, Creating a Standard Environment at *http://aka.ms/CreatingStandardEnvironments*.

Information about the Brian Keller VM is *http://aka.ms/VS11ALMVM*.

For more information about InRelease, see the website at *http://www.incyclesoftware.com/inrelease/*.

The hands-on labs that accompany this guidance are available on the Microsoft Download Center at *http://go.microsoft.com/fwlink/p/?LinkID=317536*.

2   The Beginning

Trey Research is being forced to reexamine how it creates and releases software. As with many organizations, they're thinking about change because the way they currently do things is damaging their business. They want to release reliable products quickly in response to market demand. This sounds reasonable but right now they don't have the processes and tools in place that will allow them to accomplish this goal.

This chapter focuses on ways to analyze how your business creates and delivers software. Some companies may never have clearly formulated how an idea is transformed into a product or service that's released to customers.

Of course, it's entirely possible to move from confusion to clarity. One way to begin sorting out a development process requires nothing more than a wall, sticky notes, and input from everyone involved. This is how the Trey Research team starts.

At this point, Jin is starting to wonder if he's made a big mistake.

Thursday, July 11, 2013

Complete chaos? Lost at sea? Those are nice ways to describe what's going on here. There's enough work to keep us all busy for months but no one has a good reason for doing most of it. What's worse is that no one even knows why some of the work is even there. On top of that, no one agrees with anyone else about what's important. Everyone has their own agenda. Maybe I should quit, or consider another, easier job, like herding cats, or maybe I should just look at it all as a challenge. No pain, no gain, right? I think I'm going to take all those sticky notes and put them into TFS so we have an actual backlog.

Here's the result of Jin's efforts. This is the backlog for the Trey Research application. It shows all the tasks that still need to be done.

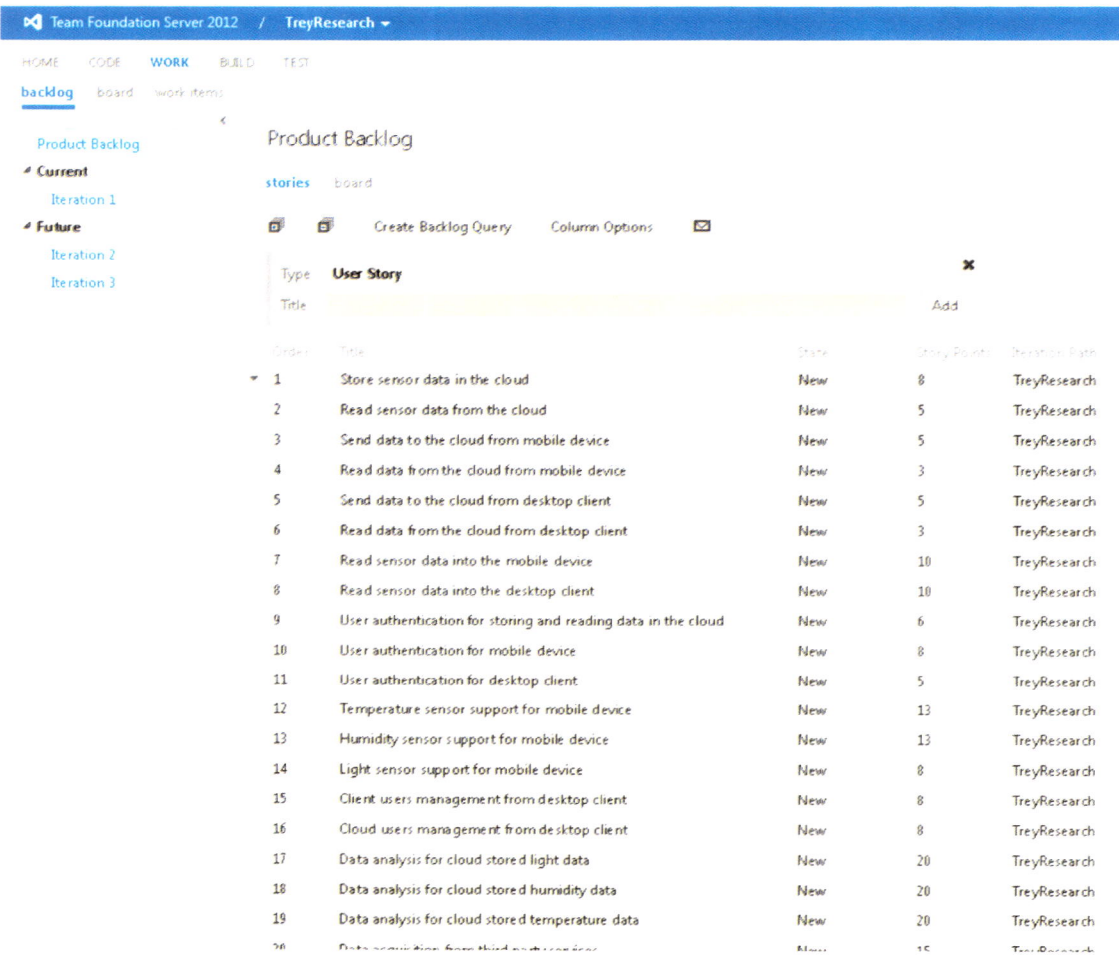

After the dust settles and they take a look at the backlog, the team discovers that there really is some order under all the seeming chaos. They find that there are eight stages to the Trey Research development process. Here are the major stages the Trey Research team identified.

1. **Assess**. Someone identifies something that could cause a change to the software. For example, an end user might make a suggestion, the marketing department might make a request after they've done some research, or the idea could come from anyone on the team. Some reasons to make the change might be to take advantage of a new opportunity, fill a business need, or fix a bug. Basically, anything that affects the code needs to be assessed to see if it's a reasonable request.

2. **Approve**. A product owner makes revisions and approves the change.

3. **Specify**. Someone analyzes the change and writes a high-level specification. There might be an associated user story or a bug report.

4. **Plan**. The team that's going to implement the change decides on its priority, estimates how long it will take to write it, and adds it to their schedule.

5. **Build**. The team writes the code, builds it, and does the basic build, unit, and commit tests. Once any bugs are fixed they can move on to the next step.

6. **Deploy**. The team deploys the new version of the software to a staging environment.

7. **Test**. The team performs more functional tests and fixes the bugs.

8. **Release**. The team releases the software to the production environment. If any problems occur here, they need to be fixed.

The team summarizes this information by drawing a *value stream map*. The map represents the flow of requests through the business and relates it to efficiency and wait time. Efficiency is the amount of time when value is being added. Wait time is a delay that adds no value. Here's the value stream map for Trey Research as it exists today.

Current Value Stream Map

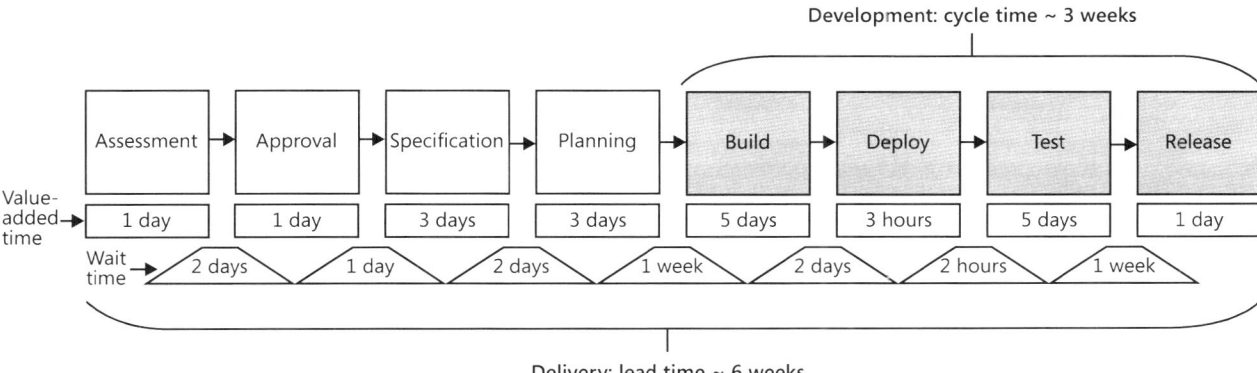

As you can see, it takes about nine weeks for a single feature to move from being assessed to being released. The metric for how long it takes for the entire process is the *lead time*. More than half of that time is spent moving from build to release. The build phase alone takes two weeks. So, not only is the process flawed because it allows unreliable software to be released, but it's slow. It takes more than a month to discover all the errors.

Next, the team decides to look at each of the eight phases they identified and to list all the steps that comprise each one. To do this, they get a lot more sticky notes, go back to the wall, mark off eight columns, and then add a few more columns just in case they missed something. Here's the final version.

The number in each column represents the work-in-progress (WIP) limit. It's the most items the column can hold. The WIP number will be adjusted later as a result of what they learn as they actually do some work. They've also added some buffer columns to designate the flow of work between the stages, still with respect to the WIP limit. When the work is likely to be blocked by external reasons they use one of the Ready for... buffer columns. They use a Done buffer column if they can remove the blocking factor by themselves.

Some people call such a display a *kanban board*. The word kanban is Japanese and means signboard or billboard. Toyota began using kanban in the 1940s. It developed the idea by watching how supermarkets stocked their shelves. Toyota applied the same principles to its factories. Beginning with an agreed upon number of physical cards, it attached one card to each new piece of work. The card stayed with the work as it progressed through the factory. New work could start only if a card was available. If there were no more cards, the work waited in a queue. When one piece of work was finished, its card was attached to a piece of work in that queue. Kanban is a pull system because work is brought into the system only when there's capacity to handle it. If the predetermined number of cards is set correctly, then the system can't be overloaded.

Kanban and kanban boards can be easily adapted to other industries, including software development. For example, practitioners of lean software development and Agile often use kanban boards to make the organization and status of their projects visible to all team members. A kanban board is useful no matter the methodology the team actually follows. For example, the kanban board we've just shown you could be adapted to work for a team that uses Scrum. The team could certainly keep the Specification, Planning, Development, Deployment, and Test columns. Even the Release column could be included if they want to introduce some DevOps ideas into their work. The column could represent developers and operations people working together to release the software. A Scrum team might change the columns to better reflect the nonsequential Scrum approach, or even have a different board for the sprint backlog. Still, the kanban board would be useful for visualizing the entire project, end to end.

Here's Jin's reactions to what's been happening.

Friday, July 12, 2013
The value steam map helped, and the Kanban board looks good, too. At least we can all see what's going on, and that's a good first step. I'm going to do the same thing for the Kanban board as I did for the sticky notes. I'll put them into TFS. I'm hoping I can convince everyone else to help me keep the board updated. Maybe when they start to see some progress, they'll be more willing to join in.

Here's the Kanban board that Jin created in TFS. The fractional numbers in the column headings (3/3, for example), show each column's WIP limit. For each backlog item in the first column, you can see all the tasks (the blue boxes) that comprise the item. Each task's status is shown by the column it's in.

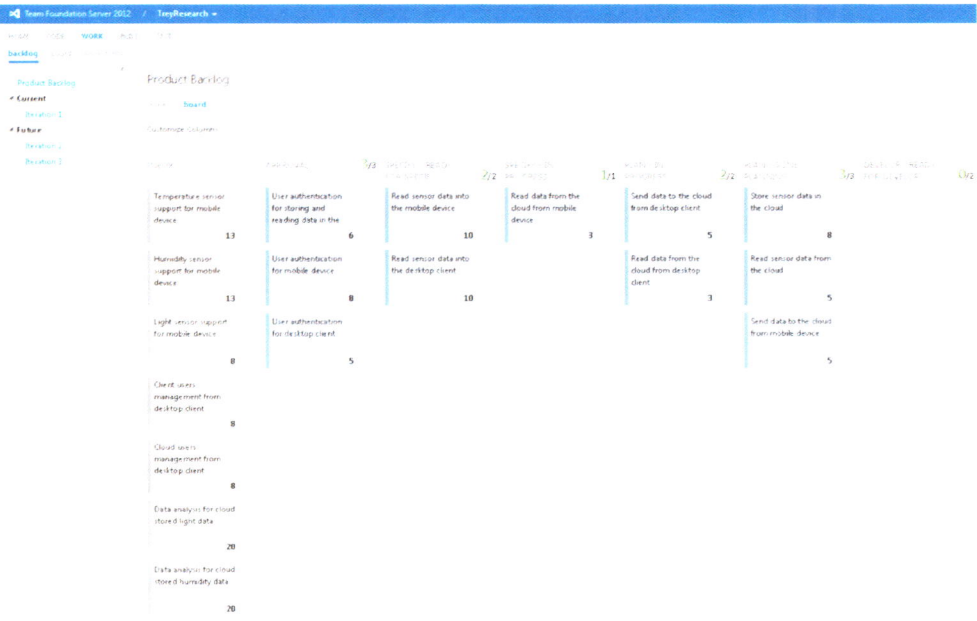

Jin however, is nervous.

> **Friday, July 12, 2013**
>
> On Monday, we start the next iteration, and it's a three week sprint. That's huge, by my standards, but it's the best I could get. Their first release took two months! I hope the work we've done so far will help us to stay focused and meet our goals.

When the rest of the Trey Research team members look at their Kanban board, they're more hopeful. They realize that they can make real improvements to the last four stages of the value stream. These stages comprise the release pipeline.

WHAT IS A RELEASE PIPELINE?

A release pipeline is an implementation of the process a business follows to create and deliver software. The pipeline begins at the point when any change (for example, a new feature or a fix) is introduced into the code base, and ends when the change is delivered to users. As the change moves through the pipeline, it undergoes a series of activities, each of which helps insure that the final software is production ready and conforms to the specified requirements. A release pipeline is automated to some degree. Automation means that (some) activities that occur in the pipeline and (some) movement from one end of the pipeline to the other happen without human intervention. A common example of automation is continuous integration, where builds are automatically triggered by check-ins and some set of tests run after the build is complete. While you may not be able (or want) to automate the entire pipeline, automating as much as is feasible can greatly improve how you deliver software.

What Are Release Pipeline Stages?

Stages are the building blocks that form a pipeline. Each stage contains a set of related steps (or activities) that occur during the software development process. These stages are in place to ensure that any change introduced into the pipeline will be good enough to be released into production. Commonly used pipeline stages are the commit stage, the automated acceptance test stage, and the release stage. Any pipeline stage can either pass or fail. If it fails, the pipeline should completely stop until the problem is resolved.

What Are Release Pipeline Steps?

A step is an action that's contained within a stage. Stages are typically comprised of several steps. The build stage, for example, encapsulates steps such as creating a build, running tests, and performing code analysis. Steps can be either automated or manual. There are many different steps that can occur in each pipeline stage. A few examples are building the software, running unit tests, provisioning an environment, running automated acceptance tests, and manually performing exploratory testing.

What Are Release Pipeline Instances, Changes, and Versions?

An instance of a pipeline is an actual occurrence of that pipeline. When a pipeline is triggered by an event (typically, when there is a check-in to the version control system) a pipeline instance is created. Each instance should be uniquely identified. The best way to do this is to name the pipeline after the check-in that triggered it. You could, for example, use the version control changeset or the version number of the code. You can use this identifier to track the stages that run inside the pipeline instance and to discover their outcomes.

Changes and versions are used interchangeably in the context of continuous delivery, and in this guidance we use both. A change or a version is some modification to the code or supporting artifacts that is checked in to version control. From the perspective of continuous delivery, every version is a candidate for release into production. Any instance of the pipeline validates only one version of the software.

What Tools Does a Release Pipeline Require?

You will use a variety of tools to create and run a release pipeline. Here's a high-level overview of the sorts of tools you'll need.

- **Tools that support the pipeline.** These tools run the pipeline, define and trigger stages, orchestrate the overall process and gather information for metrics. Specific examples of this type of tool are Team Foundation Server (TFS), Jenkins, and Go.
- **Tools that support pipeline steps.** These tools set up and run specific steps inside the pipeline stages. They can include scripting and automation languages and technologies, tools to set up and run environments, and tools to build and package software. Specific examples include PowerShell, MSBuild, MSDeploy, NAnt, System Center Virtual Machine Manager (SCVMM), and Lab Management.
- **Other supporting tools.** These tools include, for example, tools for software and script development, tools for testing and debugging, integrated development environments, tools for database management, and tools for version control. Specific examples include Microsoft Visual Studio, Microsoft Test Manager, Microsoft SQL Server Management Studio, TFS, and Visual Studio Profiler.

WHAT STAGES DO I NEED IN A RELEASE PIPELINE?

How you decide what stages your own release pipeline needs is determined by your business requirements. For example, if your business runs a website with millions of simultaneous users, you would certainly want to include a stage that tests for capacity. In many cases, the minimum set of stages required to ensure that the software is production ready, at least from a functional perspective, would be a commit stage, an acceptance test stage, a UAT stage, and a release stage.

ANALYZING THE TREY RESEARCH RELEASE PIPELINE

After they formalized the value stream map, the Trey Research team decides to take a close look at its last four stages–the build, deploy, test, and release stages that actually produce the software. These four stages are what comprise the Trey Research release pipeline. In the end, the team wants a diagram that completely defines it. There are many questions to be answered before they understand how their pipeline works. Here are some of them.

- What's in the version control system?
- How many code branches are there?
- What happens in each stage of the pipeline?
- How does code move from one stage to another?
- How many builds are there?
- How are the builds defined?
- Where are code artifacts stored?
- How many environments are there?
- How are environments provisioned?

After a lot of work they came up with a diagram. Here's what they currently have.

Current Trey Research Release Pipeline

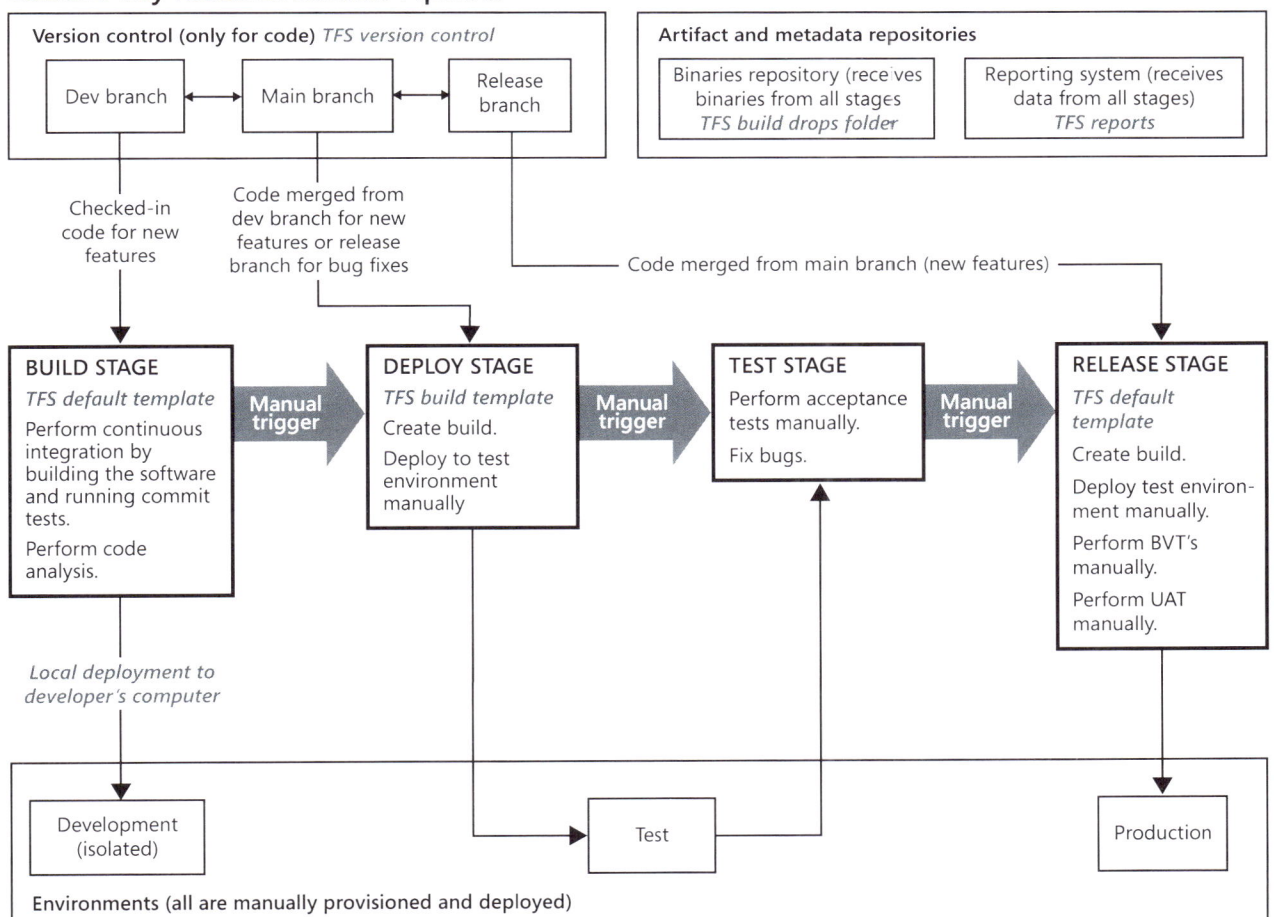

Let's walk through their pipeline to see how it works.

THE VERSION CONTROL SYSTEM

Trey Research uses TFS as its versioning system. It contains all the application code but no other artifacts, such as configuration files or scripts.

The main branch of the versioning system holds the code that's considered stable enough for the next version of the application. Iselda, the tester, performs functional tests on this code to see if it performs as it should. All new features are developed in the dev branch.

In terms of releasing code, they know that it's important to always be able to identify the code that produced the binaries currently in production. They want to ensure that any bugs found in the production code can be fixed without introducing untested features or unknown problems. To be certain that they always have the code that generated the currently running binaries, they store it in the release branch. They only need one release branch because there's only a single version of the application in production at any time. They update this branch after each release by merging the code stored in the main branch and overwriting what's in the release branch.

THE RELEASE PIPELINE

The Trey Research release pipeline has four stages: build, deploy, test, and release. The stages are sequential so every stage (except the first) depends on the previous stage to complete before it can do its job. After one stage finishes successfully, any team member can manually start the next stage. Stages themselves are composed of various steps, such as "perform continuous integration," and "perform code analysis." The team uses TFS in the build, deploy, and release stages.

The Build Stage

In this book, we make a distinction between creating a build and running tests. (Some people combine the notion of building and the notion of testing under the single term "build.") Creating a build means to prepare all the artifacts that will be run by the steps included in a stage. In other words, a build does more than compile the source code. It can also include activities such as copying files, references and dependencies, (these include supporting files such as images and software packages), and signing assemblies. Unit tests are run after the build is available.

At Trey Research, the build stage uses the code that's stored in the development branch. The build stage is defined by a TFS build definition that's based on the default template. The build definition includes information such as what should be built, when a build should run (that is, what triggers a build) the workspace, and the build controller you want to use. Here's an illustration of how a build occurs.

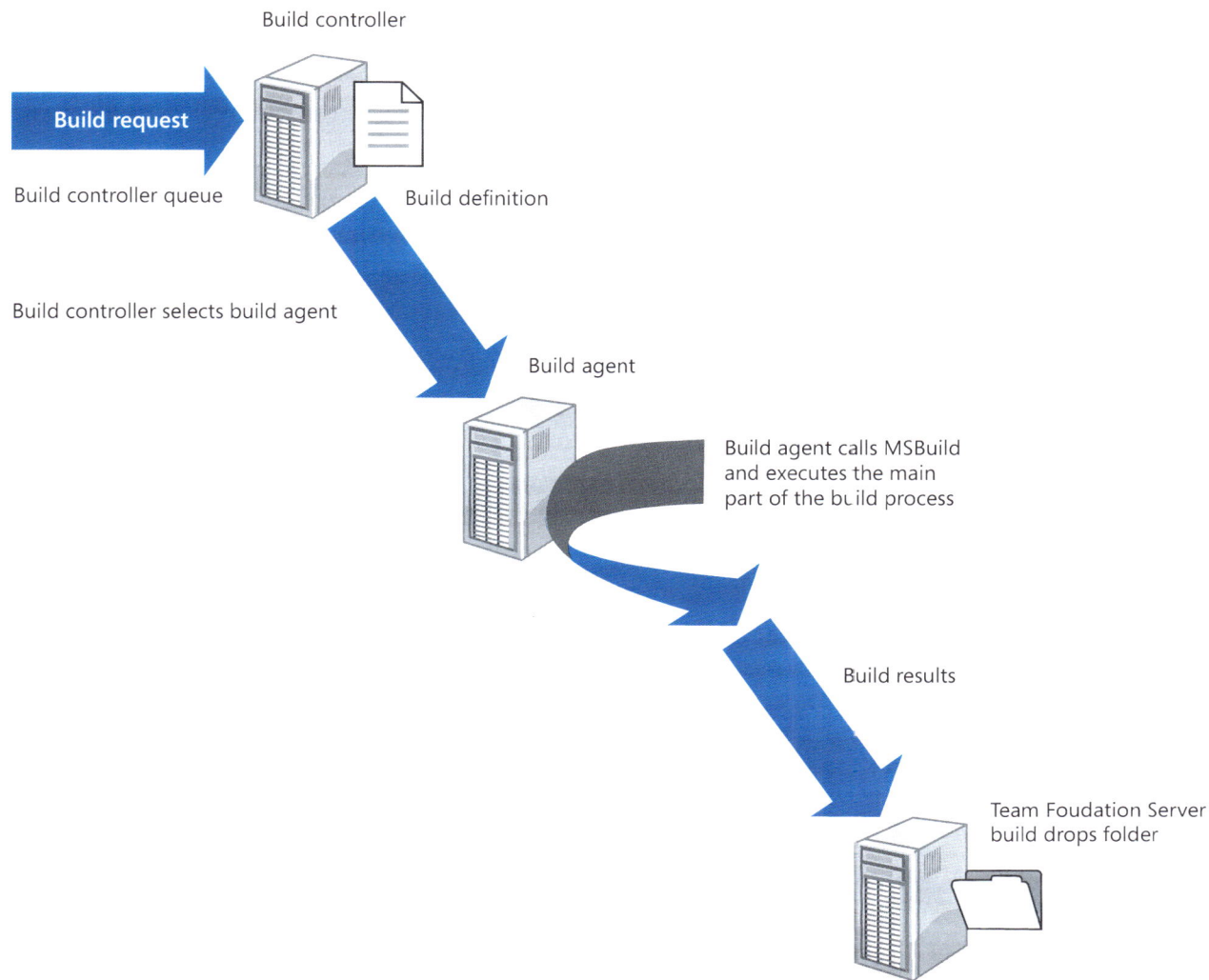

When it's determined that a build should occur (this is a part of the build definition) a team project sends a request to the build controller's message queue. (The build controller accepts build requests from any team project in a specified team project collection.) The controller retrieves the build definition and executes it.

Along with processing workflows, controllers perform some other work such as determining the name of the build and reporting the build status. They also select and manage the services of the build agents.

The build agent actually performs the build. It does all the processor-intensive work. It calls MSBuild to compile the code and run the tests. After the build agent successfully completes its build, it archives the build results to a file share called the drop location.

Trey Research's Build Definition

Trey Research uses Team Foundation build definitions that are based on the default template. Build definitions are created from within Visual Studio. If you want to recreate the Trey Research build definition, see *exercise 4 of Lab 1: Creating the Trey Research Environment.*

Before a build happens, Paulus, the developer, brings up the application on his development machine and performs some basic tests. He does this because it's easier to run the application in Windows Azure and Windows Phone 8 emulators than to deploy it to the real platform. Once he feels confident about the code, he checks it in to the dev branch, which triggers the build stage. Team members have set a TFS alert and they receive emails when a build stage ends.

The Deploy Stage

After all the steps in the build stage are complete, one of the team members uses Source Control Explorer to merge code from the corresponding development branch to the main branch. The transition from the build to the deploy stage is manually triggered. It only runs when someone explicitly requests it, perhaps through Visual Studio or the command line.

The deploy stage's job is to send the build artifacts to the test stage. However, it doesn't use the artifacts that were generated by the build stage. Instead, it retrieves the merged code from the main branch of the versioning system. It then performs another build. This stage also uses a Team Foundation build definition that's based on the default template.

The Test Stage

The test stage has its own environment. It uses the binaries and artifacts created in the deploy stage. Right now, there are only some manual acceptance tests, which aren't closely tied to user stories or requirements. There's also no dedicated tool for managing and executing test plans. Instead, Iselda, the tester, uses Microsoft Excel spreadsheets to keep track of what's going on. Moving from the deploy stage to the test stage requires a manual trigger. Iselda connects to the test environment and then runs all her tests manually.

One known issue with the test stage is that there's no UAT environment. Iselda wants one but Trey Research is on a tight budget and environments are expensive to set up and maintain.

The Release Stage

After the test stage is complete and the bugs are fixed, one of the team members uses Source Control Explorer to merge code from the main branch to the release branch. The transition from the test stage to the release stage is manually triggered. The release stage builds the code that was merged from the main branch to the release branch. Again, there are no tests to confirm that the build was successful. After the code is built, it's manually deployed to the production environment where manual smoke tests are performed as well as user acceptance tests (UAT). The release stage also uses a Team Foundation Build definition based on the default template.

The Environments

Trey Research currently has three environments: development, test, and production. Each of them is configured manually. There is no automated configuration. Each environment has a different URL that the Windows Phone 8 app uses to connect to the appropriate Windows Azure service.

THE PROBLEMS

Finally, after analyzing the pipeline, the team begins to list the problems they're having. Right now, there are so many that it's overwhelming. Here are just some of them.

- It's not clear which binary is deployed to each environment, or what source code generated it.
- There's nothing that defines the quality of the release. No one knows if it meets customer expectations or includes the agreed upon functionality.
- There are versioning inconsistencies. Incompatible versions are deployed across different environments and devices.
- There's a lot of time wasted, confusion, and errors introduced by manually moving from one stage to another.
- There are many unexpected bugs because the pipeline builds and deploys different binaries for the same version or change. These bugs make it difficult to release quickly.
- The deployment process is slow and error prone.
- They don't always work on the correct version of the application so they find and work on bugs they've already fixed.
- Regression testing is slow and error prone so they don't do it as often as they should.

Everyone is back to feeling overwhelmed.

Jin's discouraged, not just because of the chaotic state of the pipeline, but because of everyone's attitude.

Monday, July 15, 2013

After a lot of arguing, we did set up a backlog but what a struggle! Iselda wants to test everything after the iteration finishes. She says she has lots of other work to do but Zachary made it clear that the Trey Research app has top priority. She finally came around and is going to test along with development and not wait until they finish. Raymond is really difficult. He didn't understand why he had to come to our meetings when there are already plenty of problems in the server room. He left as soon as he could. We did manage to add a couple deployment tasks for him to the backlog.

I told everyone I'd take this iteration's backlog and all the tasks and put them into TFS. Paulus said (and I know everyone else was thinking the same thing) that it's a waste of time and that we should be coding. They'll see.

Here's the result of Jin's work. The user stories are in the first column. The blue boxes are the tasks that make up each user story.

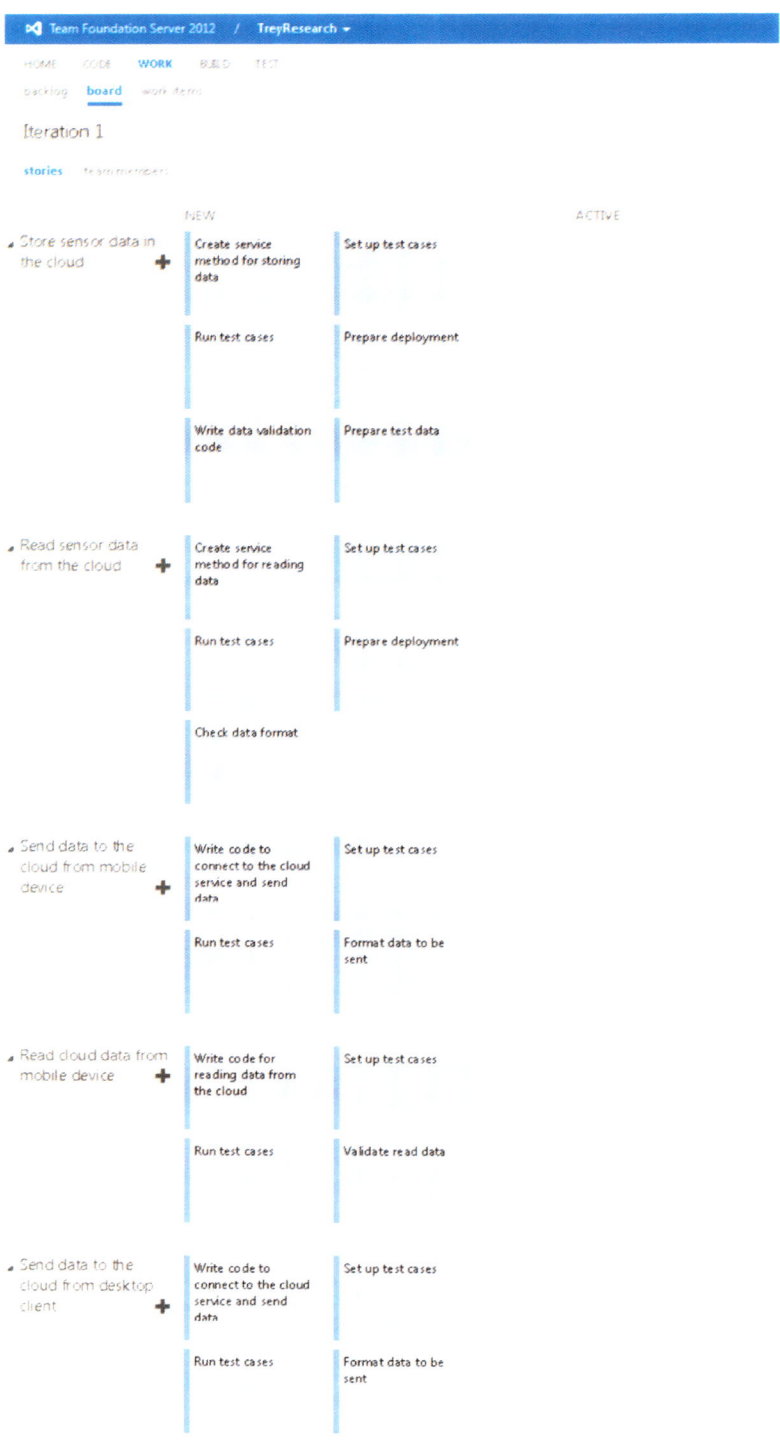

Still, even with all the preparation, there are many problems. By the middle of the iteration, Jin sees trouble.

Wednesday, July 24, 2013

It's clear we won't reach our sprint goal by August 2nd. That's what the iteration board and the burndown chart tell us. One small victory is that Paulus admitted that it's useful to keep information in TFS but none of them help keep the board up to date. At least we know that whatever's going to be delivered is actually finished, or "done-done" as some folks say. Of course, that's from the developer's side. I still don't know how we're going to coordinate with Raymond to get our app into the production environment and out to our users.

Here's the latest iteration board. It shows the user stories in the first column. The blue boxes show the tasks that comprise each story. As you can see, there are only a few tasks in the Closed column. All the others are either in the New column, which means that they haven't been started or in the Active column, which means that they're not finished. Because the team is already in the middle of the iteration, it seems clear that they won't meet their goal.

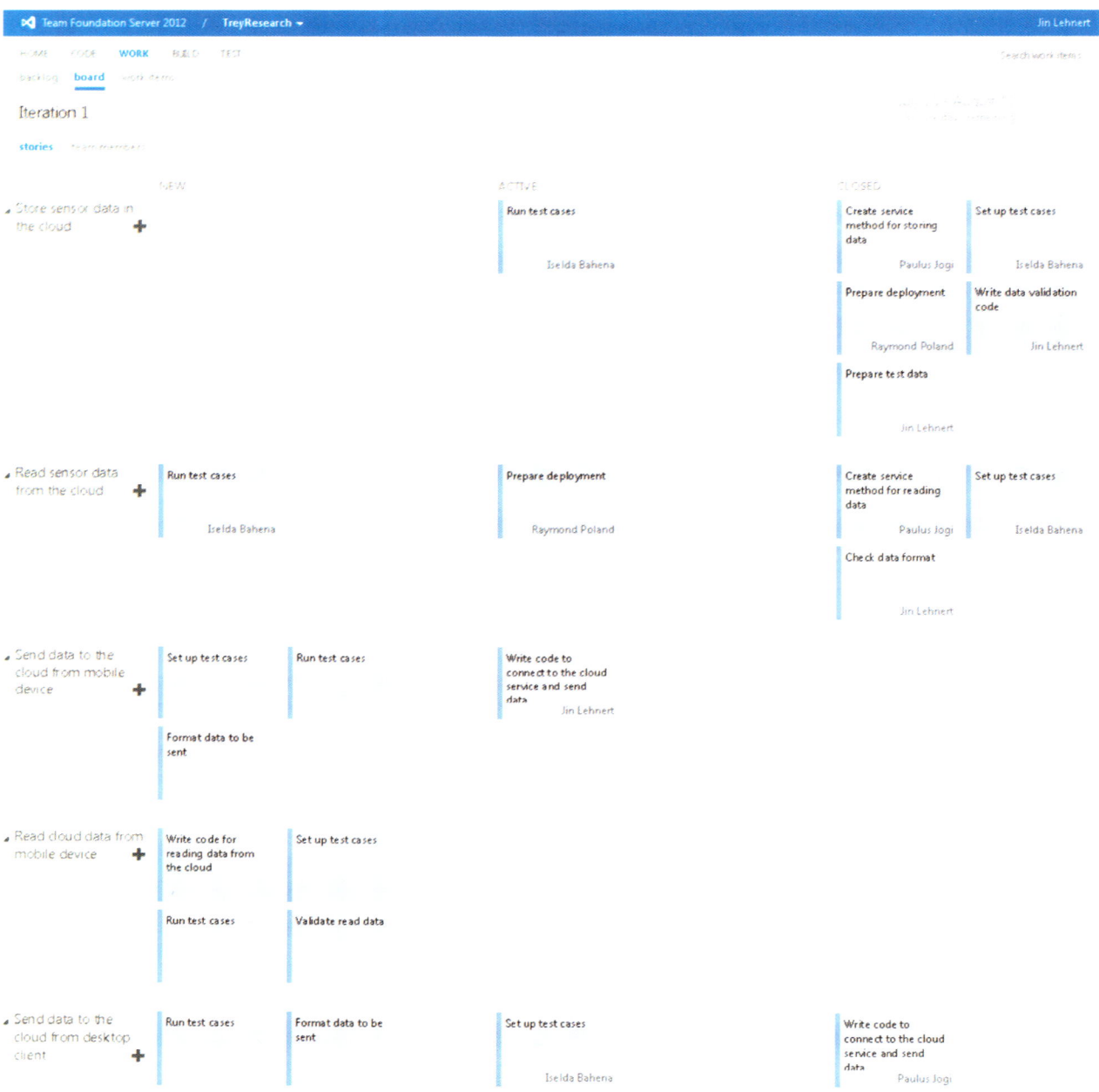

Jin sees even more bad news.

> **Wednesday, July 24, 2013**
> We've ignored the WIP limits in our buffer column for development. Nobody was concerned or saw it as a problem. They didn't seem to mind at all. Maybe they think it shows how hard they can all work. Right now, everyone's focused on producing a release, but we really need to change the way we do things around here.

Here's what the product backlog shows. There are four items in the Develop—Ready For Develop column when there should only be two.

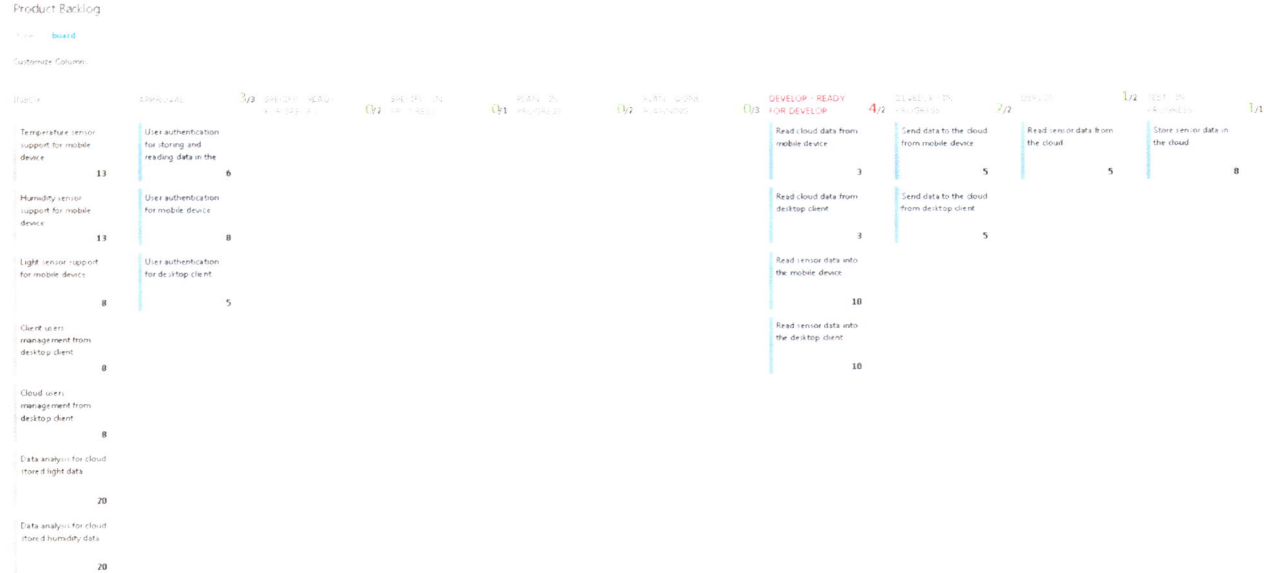

At the end of the iteration, the news still isn't good.

> **Friday, August 2, 2013**
> We completely missed our forecast. We did manage to deliver a working version of the basic services, along with a simple desktop client. But we're missing many features we planned for this milestone, and I don't think anyone will buy what we have now. There were even some features that we developed that weren't delivered to customers. They stayed in the Release column of our board, waiting for the binaries to be promoted to production. I won't blame Raymond, because this is the whole team's fault. But I do say he's a part of the team, no matter how upset he gets when I mention it. If we don't all work together, we'll never get rid of the bottleneck just before the release column, where items keep piling up.

Here's what the Kanban board looks like at the end of the iteration (see page 34 for an expanded view):

Here are Jin's final comments for this iteration.

Friday, August 2, 2013
Zachary's really angry. It took almost five weeks to get that little green square at the end of the board, and it's all alone. It doesn't matter if there are four more items almost there. If they're not deployed to production, nobody cares. Stakeholders won't support this much longer. The one thing Zachary did like was that he could see the status of every work item. He said that now he really knew what was going on. He liked it so much that he told the rest of the team to help me keep it up to date. Everyone looked at their shoes.

SUMMARY

This chapter talks about ways to understand your development process, from the time you have only an idea for a feature or product, until the time your idea is realized and you can deliver it to your customers. Tools aren't what matter at this stage. What matters most is that everyone involved have a voice and participate in the analysis. The procedure we use has five phases.

1. Start with the stated goal of understanding how you develop your software. You only need some sticky notes, a wall, and your team members. Let everyone participate, adding steps that they think are important. Try to identify the most general activities.

2. Once you've got a sense of the major activities that happen (for example, in this chapter, Trey Research narrows it down to eight), formalize your system with a value stream map. Using the map, relate the activities to how much time you spend accomplishing them and to how much time you're blocked.

3. You might want to create a Kanban board to get a more granular understanding of what happens during each of the major stages you've identified.

4. Focus on the release pipeline itself. Draw a diagram that shows how the bits flow from one stage to another. List the steps in each stage, the tools you use, the environments each stage depends on, and how you store and keep track of builds and all other artifacts that affect your development process.

5. List the problems you're having.

6. (Optional) Take a deep breath.

We also talked about release pipelines. A release pipeline is an implementation of the process a business follows to create and deliver software. It consists of stages. Stages are the building blocks that form a pipeline. Each stage contains a set of related steps that occur during the software development process. Steps are actions that are contained in stages. A new version triggers an instance of a pipeline. A version is some modification to the code or supporting artifacts that is checked in to version control.

For the Trey Research team, life is still very difficult. They've made progress in terms of using TFS to help them organize their work and to gain visibility into their release process. However, they're still missing major goals and not shipping what they should.

WHAT'S NEXT

In the next chapter, the Trey Research team, although initially overwhelmed, starts to put the pieces together. First, they figure out how to set some priorities. They also start to think about orchestration. By orchestration we mean the arrangement, coordination, and management of the release pipeline. The team makes a decision and decides they're going to gradually move toward a continuous delivery pipeline. They'll build it using the TFS default build template and the Lab Management default template, along with some supporting tools.

MORE INFORMATION

There are a number of resources listed in text throughout the book. These resources will provide additional background, bring you up to speed on various technologies, and so forth. For your convenience, there is a bibliography online that contains all the links so that these resources are just a click away. You can find the bibliography at: *http://msdn.microsoft.com/library/dn449954.aspx*.

The ALM Rangers have put together a collection of guidance and labs to help you use TFS for application lifecycle management. For more information, go to Visual Studio Team Foundation Server Requirements Engineering Guidance at *http://vsartfsreguide.codeplex.com/*.

The ALM Rangers have put together a collection of guidance and labs to help you understand how to customize and deploy Team Foundation builds. For more information, go to Team Foundation Build Customization Guides at *http://vsarbuildguide.codeplex.com/*.

To understand how to use the TFS Kanban board to understand the status of your project, see Work on the Kanban board at *http://tfs.visualstudio.com/learn/work-on-your-kanban-board*.

To learn how to customize the TFS Kanban board, see Kanban customizable columns, "under the hood" at *http://blogs.msdn.com/b/visualstudioalm/archive/2013/03/04/kanban-customizable-columns-under-the-hood.aspx*.

The hands-on labs that accompany this guidance are available on the Microsoft Download Center at *http://go.microsoft.com/fwlink/p/?LinkID=317536*.

Expanded view of the Kanban board at the end of the iteration:

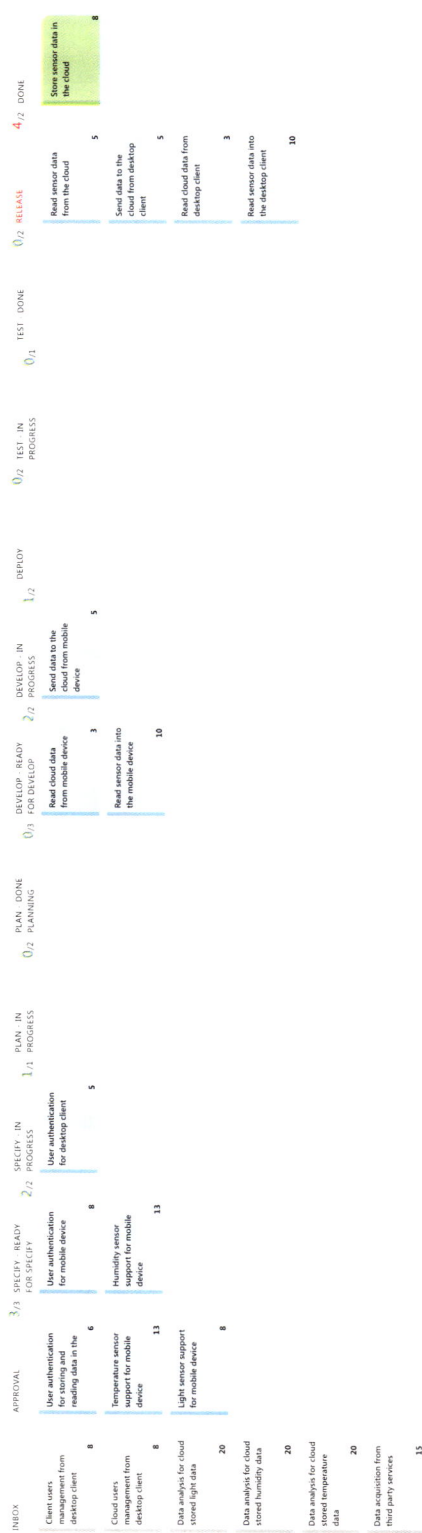

3

Orchestrating the Release Pipeline

In the last chapter, the Trey Research team analyzed their development process and their pipeline. When they were done, they listed all the problems they were having. It was a long list. In this chapter they decide on what the most important problems are and what improvements they can make to their pipeline to address them. We'll see that they are going to *orchestrate* the pipeline, which means they are going to add the capabilities to arrange, coordinate, and manage it. Orchestration can occur at both the pipeline level and at the stage level. To introduce you to the topic of orchestration, we'll first discuss some general principles to follow. While much of this guidance is true for any release pipeline, the intent in this guidance is to eventually create a continuous delivery pipeline. Later in the chapter, to give you a concrete example of how to put those principles into practice, we show the improvements the Trey Research team implements for their pipeline by using the default Lab Management and TFS build templates. The changes they make in this chapter are the foundation for other changes they will make later on.

GENERAL PRINCIPLES FOR ORCHESTRATING A PIPELINE

This section discusses general principles you can use as a guide when you orchestrate a pipeline.

Make the Pipeline Complete

Incorporate as many of the activities that comprise your development and release processes as possible into the pipeline as steps. Where you can, make the steps automatic rather than manual.

Translating activities into steps defines your process and makes it both visible and predictable. You'll be better able to identify areas that can be improved. Also, a well-understood process makes it easier to predict how long it takes and how much it costs to put a new version of the software into production.

If you're interested in adopting some of the principles that are part of the DevOps mindset, a complete pipeline can help you. The pipeline emphasizes the performance of the entire system and not the performance of a single department or silo. Building a complete pipeline can encourage collaboration between groups that may not normally communicate with each other.

Use Automation When Possible

When you can, trigger stages automatically rather than manually. Of course, there are times when you must use a manual trigger. For example, a stage may contain manual steps. There may even be cases where you want to manually trigger stages that have only automatic steps. The most common example is a release stage, where you want control over when the version is released and to what production environment.

Automation generally makes the pipeline faster because you're not waiting for someone to manually trigger the stages. Automated stages also have a near zero, fixed cost. Increased speed and reduced costs make it more practical to have smaller, more frequent versions of the software, which are themselves easier to debug and release.

Move Versions Forward in a Continuous Delivery Pipeline

After a check-in triggers the creation of a pipeline instance and begins to be propagated through its stages, there's no way back. You only move forward through a continuous delivery pipeline, never backwards. In other words, in a continuous delivery pipeline, the same version never travels through the same stage in the same pipeline instance more than once.

Moving in only one direction helps ensure the reliability of the software. If there's a failure the pipeline stops. A new check-in that addresses the problem is treated like any other version, runs through the entire pipeline, and must pass all the validation stages.

TRIGGER THE PIPELINE ON SMALL CHECK-INS

Small, significant check-ins should go through the pipeline as soon as possible. Ideally, every single check-in to version control should trigger the pipeline. Also, check-ins should be done frequently, so that a new version of the software differs only slightly from the previous version.

There are multiple benefits to working with small, frequent check-ins. If a small change makes the pipeline fail, it's very easy to figure out what the problem is. You also reduce the batch size of items that you deploy and test over the different environments. Ideally, the batch size should be a single item. Additionally, propagating small changes reduces work queue lengths (with their associated hidden costs), because small changes go through the pipeline much faster than large ones. Small changes also mean that you get fast feedback. In other words, fail fast and fail often.

Keep the Number of Pipeline Instances Small

Keep the number of running pipeline instances small. You want to focus on small sets of changes and not be distracted by constantly switching from one version to another.

Remember the kanban approach that limits the number of items that constitute work in progress (WIP). You'll need to find a balance between the WIP items, and the time that is spent, on average, to close each item. Also, fixed limits are imposed by the availability of people and resources such as environments.

If your queue length and WIP are increasing, temporarily reduce or even stop check-ins. Dedicate more people and resources to moving the WIP items you already have through the pipeline. When the queues and WIP return to reasonable levels, resume work on new features and start checking in new versions again.

Concentrating on a limited number of versions means that you'll get them through the pipeline quickly, which reduces cycle time. You're also less likely to have one pipeline instance block progress in another because of resource conflicts. If you work to optimize your process and to remove bottlenecks, you'll find that the pipeline runs faster, and you'll be able to support more concurrent instances.

Run Stages and Steps in Parallel

Whenever you can, run stages and steps in parallel. You can do this when one stage or step doesn't rely on the results of the preceding stage or step. This guidance is particularly relevant to continuous delivery pipelines, which depend on getting fast results from each stage. Versions enter the first stage, where many problems are detected quickly. From that point on, if your initial tests are thorough, you should have a reasonable chance that the software will go through the rest of the pipeline without problems. Given this assumption, you can save time by running some stages in parallel. Running your stages and steps in parallel gives you faster feedback on your software and helps to reduce cycle time.

Don't Mistake Steps for Stages

Be careful when defining the stages of your pipeline. A stage lets a version advance until it is considered ready for production. It's generally made up of multiple steps. It's easy to confuse what is really a step for an entire stage.

Deployment is a common example of this mistake. Deployment is not a stage in and of itself because it does nothing to validate the application. Instead, think of deployment as a single step in a larger set of activities, such as testing, that comprise a stage. These activities work together to move the application towards production.

For example, if you run UATs and then deploy to another environment, the deployment is not a stage. Instead, you have a stage with manual steps (the UATs) and an automatic step (the deployment).

Orchestrating a Step

Orchestrating an individual step largely depends on what a specific step is intended to do, so it's not possible to provide a general approach. Implement the orchestration with any technology you feel is easiest to use. Examples of these technologies include Windows Workflow Foundation, PowerShell, or shell scripts. The code within the step should control the step's orchestration.

Stop the Pipeline on a Failure

If a version fails any stage, that instance of the pipeline should stop immediately. (Other running instances shouldn't be affected.) Fix the problem, check in the solution, and run the new version through a new pipeline instance. The fix should be done as soon as possible, before going on to other activities, while the issues are still fresh in everyone's mind. This guidance is particularly relevant to continuous delivery pipelines, which assume that any check-in, if it successfully goes through the pipeline, can be given to customers. If the pipeline fails, the version isn't suitable for release.

Stopping the build ensures that defects never reach production. It also engages the entire team in fixing the problem quickly, thus reinforcing the notion that everyone owns the code and so everyone must work together to solve problems. This is a situation where you can apply the DevOps mindset and get everyone involved.

Build Only Once

Build your binaries and artifacts once. There should be a single stage where the build occurs. In this guidance, we call this stage the *commit* stage. The binary should then be stored someplace that is accessible to your deployment mechanism, and your deployment mechanism should deploy this same binary to each successive environment. This guidance is particularly relevant to continuous delivery pipelines where a check-in triggers a build and that specific build goes through all the validation stages, preferably as an automated process.

Building once avoids errors. Multiple builds make it easy to make mistakes. You many end up releasing a version that isn't the version you tested. An extreme example would be to build your source code during the release stage, from a release branch, which can mean nothing was tested. Multiple builds can introduce errors for a variety of other reasons. You may end up using different environments for each build, running different versions of the compiler, having different dependencies, and so on. Another consideration is that building once is more efficient. Multiple builds can waste time and resources.

Use Environment-Agnostic Binaries

Deploy the same binaries across all the environments that the pipeline uses. This guidance is particularly useful for continuous delivery pipelines, where you want as uniform a process as possible so that you can automate the pipeline.

You may have seen examples of environment-dependent binaries in .NET projects that use different debug and release configurations for builds. During the commit stage, you can build both versions and store the debug assemblies in an artifact repository and the debug symbols in a symbol server. The release configuration should be the version deployed to all the stages after the commit stage because it is the one that's optimized. Use the debug version to solve problems you find with the software.

An interesting scenario to consider is localization. If you have a localized application and you use a different environment for each language, you could, instead, have subsets of environments, where a subset corresponds to a language. For each subset, the deployed binaries should be the same for all the environments contained within it.

Standardize Deployments

Use the same process or set of steps to deploy to each environment. Try to avoid environment-specific deployment steps. Instead, make the environments as similar as you can, especially if you can't avoid some environment-specific deployment steps. Treat environment-specific information as parameters rather than as part of the actual release pipeline. Even if you deploy manually, make your procedures identical.

Standardizing deployments offers another opportunity to apply the DevOps mindset. To successfully standardize your deployments requires close collaboration between developers and operations people.

The primary advantage of standardizing your deployment process is that you'll test the process hundreds of times across all the environments. As a result, you'll be less likely to run into problems when you deploy to your most critical environments, such as production. Standardization is an example of following the "fail fast, fail often" maxim. Test your procedures often, and begin as soon as you can so that you find problems early in the release process.

Keep Stages and Steps Source and Environment Agnostic

As discussed in "Standardize Deployments," deploying the same way to every environment enables you to separate configuration data from components that make up the actual release pipeline, such as scripts. Distinguishing data from components also makes it possible to have steps that don't need to be adapted to different environments. You can reuse them in any stage. In turn, because stages are made up of steps, the stages themselves become environment agnostic.

You can apply the same principle to sources that supply information to the steps and stages. Examples include a path name, a branch in the version control system, or an artifact repository. Treat these as parameters instead of hardcoding them into the pipeline.

Environment-agnostic stages and steps makes it much easier to point your pipeline to different sources or targets. For example, if there's an urgent issue in production that must be addressed quickly, and the production code is archived in a release branch, you can fix the problem in the release branch and then point the pipeline to that branch so that the new version can be validated. Later, when life is calmer, you can merge the fix into the main branch, point the pipeline to the main branch, and validate the integration.

By keeping configuration data separate from the steps and stages, your pipeline becomes more flexible. It becomes much easier and faster to address problems, no matter in which environment or code branch they occur.

Build Verification Tests

Run build verification tests (BVT) after every deployment to make sure that the application works at a basic level before you go on to further validations. Even manual BVTs are useful, although you should try to automate them as soon as possible. Again, this is a case of "fail fast, fail often." It saves time and money to detect breaking changes as early as you can.

BVTs are also known as smoke tests. Wikipedia has an overview of *smoke tests*.

Deploy to a Copy of Production

Have the environment where you run your key validation stages as similar to the production environment as possible. Differences between environments can mean potential failures in production. There are many ways environments can differ. For example, there may be different versions of the operating system, different patches, or different updates. These variances can cause the application in the actual production environment to fail with errors that are difficult to understand and debug. Even a good deployment script won't address these sorts of problems.

Here's another opportunity to adopt the DevOps mindset. If operations and development collaborate, it will be much easier to create accurate copies of the production environment. It's to everyone's advantage to have smooth releases to production. Bugs that appear in the production environment but that are seen nowhere else can take many long hours to find and fix, delay release schedules, and raise everyone's stress level.

Version According to the Pipeline Instance and the Source Code

Version binaries according to the pipeline instance and the source code that generated them. (Note that a pipeline instance is uniquely identified. For example, in TFS you can use a combination of numbers and characters for the name.) It then becomes easy to trace any running instance of the application back to the correct source code and pipeline instance. If there are problems in production, you'll be able to identify the correct version of the source code that contains the bugs.

For .NET Framework projects, modify the **AssemblyInfo** files that are included as properties of the project. Do this from within the pipeline as a step in the commit stage, just before the build step. The version number can be automatically generated based on criteria that fits your situation. Semantic versioning, which defines a consistent scheme of version numbering, might be one approach you could adopt. (For more information, go to *Semantic Versioning 2.0.0*.) For artifacts that are not converted to binaries, such as HTML or JavaScript code, have a step in the commit stage that embeds a comment with the version number inside the file.

Using an incorrect version of the source code can introduce regression bugs or untested features into your application. A versioning scheme that clearly relates artifacts to the pipeline instance and source code that generated them prevents this.

Use an Artifact Repository to Manage Dependencies

An artifact repository makes it easier to manage dependencies and common libraries. The standard repository for .NET Framework projects is NuGet, which is fully integrated with Visual Studio. You can easily configure it to use the package restore feature so that it can be accessed by the build step in the commit stage. For more information, see *Using NuGet without committing packages to source control*.

Using an artifact repository has the same benefits as using a version control system for your code. It makes them both versionable and traceable. There is an authoritative source where artifacts are stored. Additionally, you can also share common utilities and libraries with other teams.

TREY RESEARCH'S FIRST STEPS

Now let's take a look at what the Trey Research team is doing. When we left them, they were looking at so many problems they didn't know where to start.

Realizing that going out of business isn't the solution they're looking for, the Trey Research team decides to concentrate on the issues that are causing the most pressing business problems. Here's the list.

Issue	Cause	Solution
It's not clear which binary is deployed to each environment, or which source code generated it. Incompatible versions are deployed across different environments and devices.	The generated binaries and artifacts aren't versioned correctly.	Replace the build stage of the pipeline with a commit stage. Version the binaries and artifacts during the build. Base releases on the versions, not the deployments.
There's nothing that defines the quality of the release. No one knows if it's up to customer expectations or includes the agreed upon functionality.	There are no formal means to verify that the code is what the customers and stakeholders want.	Introduce acceptance criteria and acceptance test cases in the specification.
There's a lot of time wasted, confusion, and errors introduced by manually moving from one stage to another.	Versions are propagated manually through the stages of the pipeline.	Begin to configure the pipeline to propagate versions automatically.
Feedback about each version arrives late in the process and there isn't much information. It's difficult to improve the applications at a pace that's reasonable for the business.	There aren't enough feedback loops in place. The ones that are there don't provide useful information because there are so many different versions of the binaries. Also, because the pipeline is sequential it's slow, so what feedback there is takes a long time to be available.	Improve the feedback loop and provide better information. Relate the information to the same version across all stages. Begin to run stages in parallel where possible, so that the information arrives earlier.

Thinking about it, Jin decides to propose a plan to the team.

Sunday, August 4, 2013

I spent all weekend thinking up a strategy to get us on the right path. I know everything would be better if we worked together. I'm going to propose to Raymond that we try to improve the pipeline. He's been working all weekend trying to do a release so I think this might appeal to him. Paulus and Iselda can keep working on new features so we actually have something to deploy. I'm not sure where we should start. I need to talk it over with the team. Let's hope I can convince Zachary.

After a lot of discussion, Zachary agrees. The next step is to start redesigning the pipeline so that the team can solve their biggest problems. They know they want to build once, they want a reliable versioning system, they want some automation, they want better feedback, and they want some criteria to help them create better tests. The problem is, they're not sure what they should do to achieve these goals.

Raymond's attitude isn't as unreasonable as it may seem. Operations people are very protective of their release environments and they have good reasons for it. They're completely responsible for what happens there and many of them will not agree to automatic deployments to production. Jin will have to make some compromises.

Changing How They Work

As they think about the new pipeline, and what continuous delivery means, the team realizes that it's going to change how they work. They've already started to analyze how they create and release software, so they decide to revise their value stream map and their Kanban board. Here's what the new value stream map looks like.

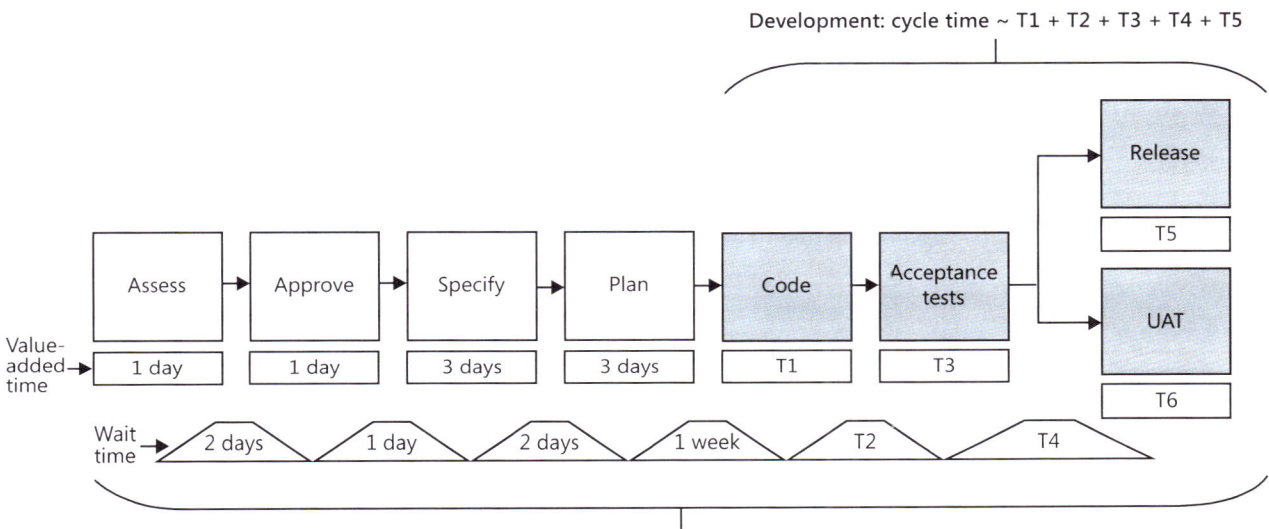

They don't know yet how long it will take them to complete each phase of the value stream map that will correspond to the new pipeline. So, for now, they're calling these times T1 through T6. After they have completed a few iterations and have some data, they'll take out these placeholders and put in actual numbers.

The revised Kanban board reflects the collaborative atmosphere they want to foster. There are two columns for the development process, Coding and Delivering. The first column represents tasks for developers, but the second represents tasks that involve the new pipeline. This column will contain items for everyone involved in delivering the software. Of course, this includes developers but it also includes testers, operations people and managers. Here's their drawing of the new board.

Inbox 5	Approve 3	Specify 3		Plan 5		Coding 4		Delivering 4	Done
		Ready for spec	In Progress	In Progress	Done	Ready for coding	In Progress		

Here are Jin's thoughts on what a new pipeline will mean to Trey Research.

Monday, August 5, 2013

A new continuous delivery pipeline means changes. We have a new value stream map and a new Kanban board. The Delivering column will contain all the items that go through the pipeline. We'll even use this organization to help us build the new pipeline, and, to prove we're serious, we've added work items to make sure we get started. Of course, now some of the work for new features is blocked and breaks the WIP limit. Let's hope we don't get too far behind. I think Raymond hates me.

Here's what the new board looks like in TFS. You can see the new items that are related to the pipeline, such as "Release pipeline configuration."

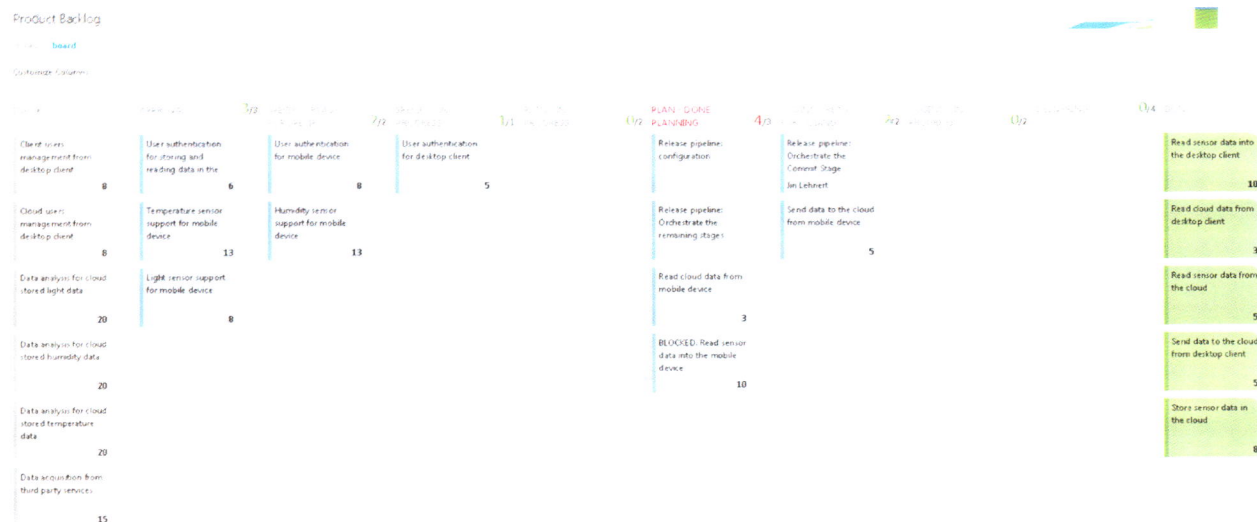

Jin has a few more comments to make about how the release process is changing.

Monday, August 5, 2013
We're shortening the iterations to two weeks. Maybe this is a bad idea. If we can't deliver in three weeks, how can we deliver in two? I'm hoping that because there's less to do, we can focus better. Also, we'll get feedback on what we're doing sooner.

Here's the latest product backlog. Notice that, even though he has strong opinions, Raymond is taking on more responsibilities and is really involved in implementing the pipeline.

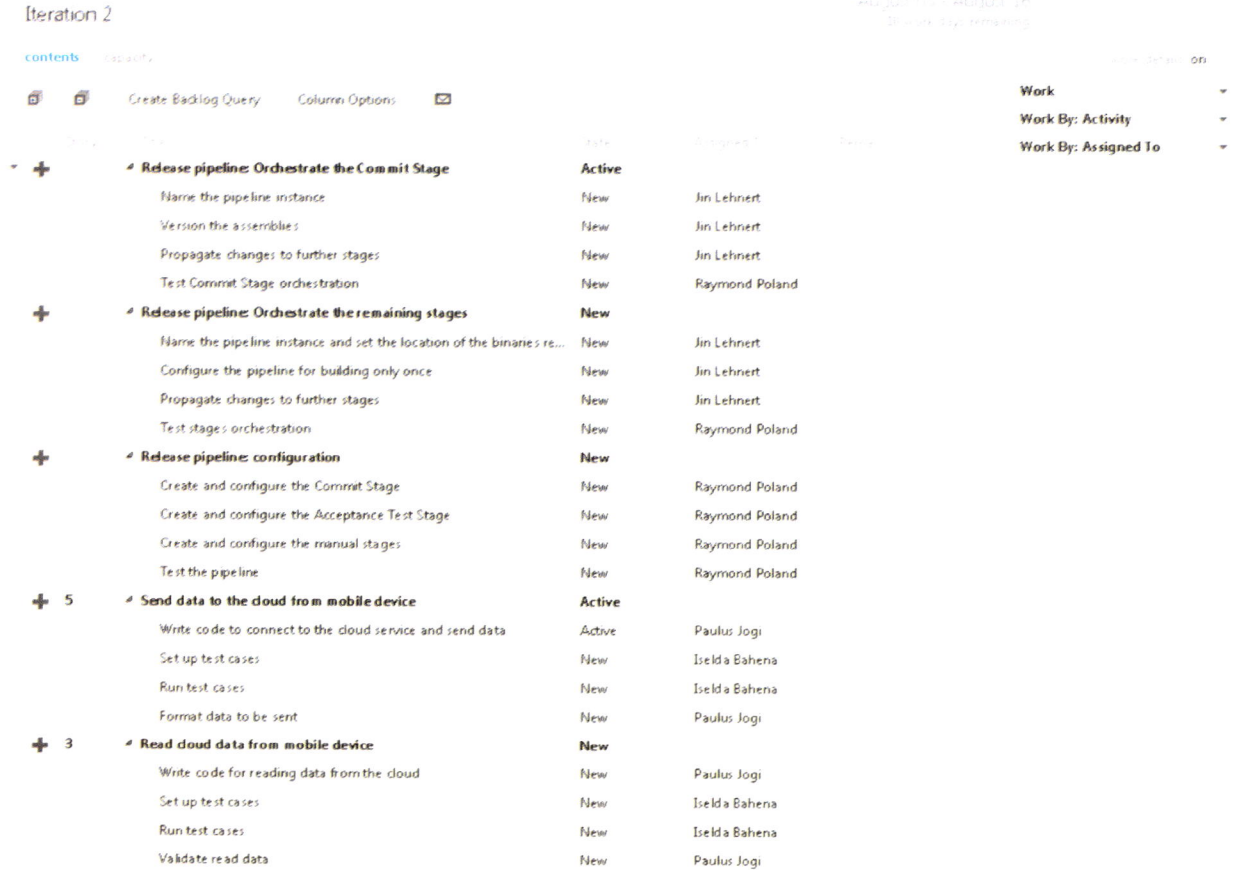

TREY RESEARCH'S PLAN FOR ORCHESTRATING THE PIPELINE

Trey Research has planned what they need to do for this iteration of the pipeline. They're going to focus on orchestration, which will lay the framework for the rest of the improvements they'll make in the future. They know that when an instance of the pipeline runs, it should follow a sequence of actions that is defined when the pipeline is configured. This sequence is the pipeline orchestration. As they designed the orchestration, they tried to keep some best practices in mind. (These principles are discussed earlier in this chapter.) In particular, they want to:

- Build once.
- Whenever possible, trigger the pipeline stages automatically rather than manually.
- If any stage fails, stop the pipeline.

Here are the actions that they've decided should occur within the orchestration.

1. Create the pipeline instance and give it an identifier based on the related version. Reserve the necessary resources.
2. Run any automatically triggered stages by passing them the appropriate parameters, as defined in the pipeline configuration. Provide a way to relate the states to the pipeline instance being run.
3. Gather all the relevant data generated at each stage and make it available for monitoring.
4. For manual stages, give people the ability to run and manage the stages within the context of the pipeline.

This diagram shows a high-level view of the Trey Research pipeline orchestration.

Pipeline-level Orchestration

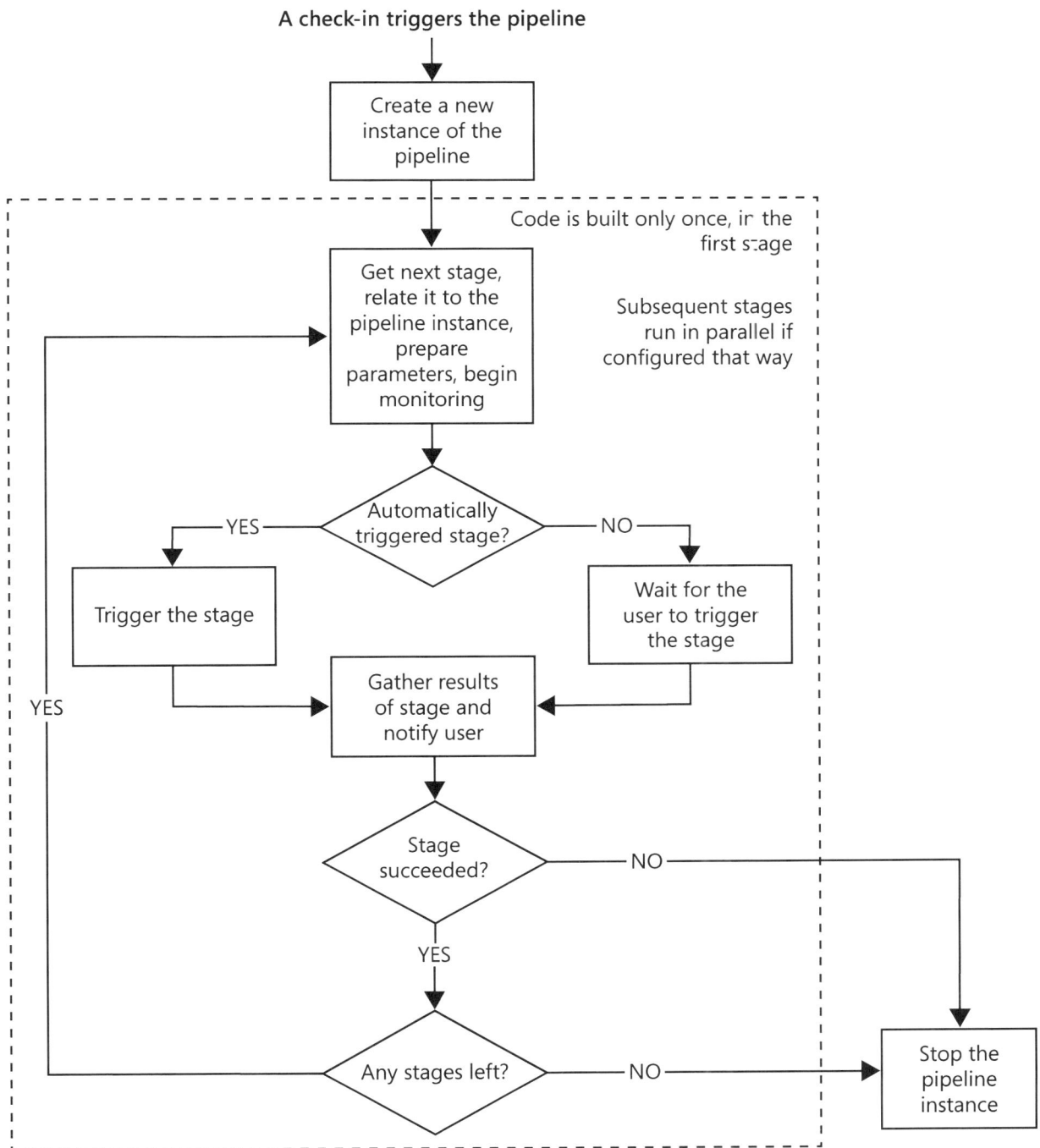

Remember that continuous delivery pipelines only move forward. After a version triggers a pipeline instance and begins moving through the stages there is no way back. No version goes through the same stage in the same instance more than once.

ORCHESTRATING THE STAGES

When you orchestrate a stage you define how it's triggered and the steps it contains. Here are some principles that the Trey Research team used to guide the design of their pipeline.

- Steps are where all the work happens. Look at every activity that occurs during the development and release process and identify the ones that can become steps in one of the stages. See if the steps can be automated.

- Implement steps so that they can be incorporated into any stage by passing the appropriate parameters.

- When you define a step, think of it as an action. Examples are "Set the pipeline instance," "Run unit tests," and "Run automated tests."

- Make sure to collect the relevant data and results for monitoring.

Here are the actions that occur within the stage orchestration.

- Provide the automated steps within a stage with the appropriate parameters.

- Make sure that any resource required by an automatic step is ready to be used when the step starts. The most common example is environments. If there is no automatic provisioning of environments in place, and the requested environment is being used by another pipeline instance, the stage orchestration should make the step wait until the environment is ready to be used again. The orchestration should maintain a queue of instances in order to make sure that access to the shared resource occurs in the correct order.

- For manual steps, wait for users to complete their tasks and then take back control when they are done.

- Make sure to collect the relevant data and results for monitoring.

The diagram shows a high-level view of the Trey Research stage orchestration.

Stage-level Orchestration

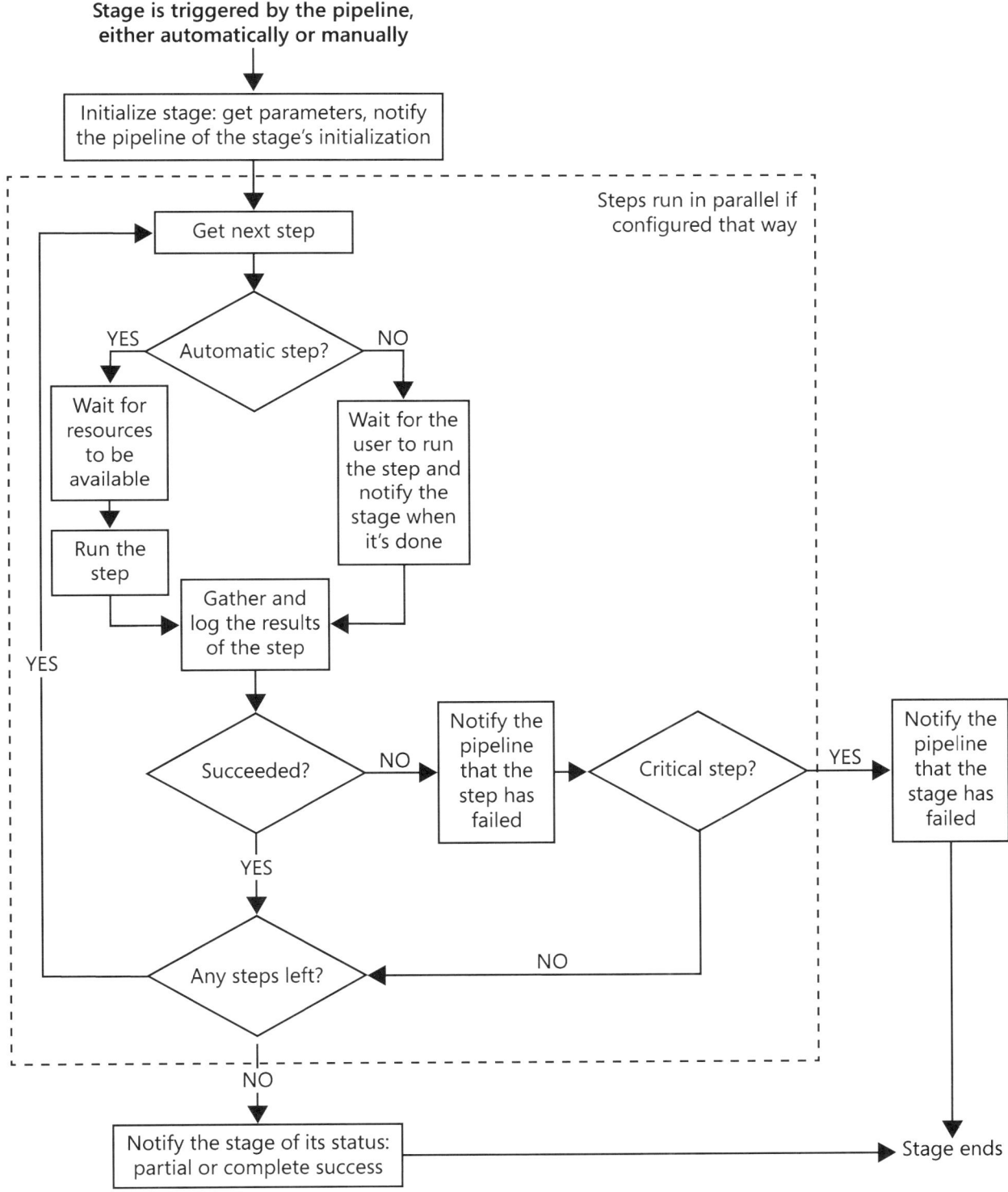

A stage can either pass or fail. If it fails the pipeline should stop and not be restarted until the problem is solved.

Building a Release Pipeline with TFS

Trey Research is going to use TFS to build their pipeline. It's a technology they already know, and they've created simple build definitions before. Although not everything necessary to create the pipeline is a standard feature in TFS, there are ways to add the required capabilities. The easiest approach is to use the default build templates provided by TFS and Lab Management and to customize them when necessary.

...w feature of Lab Management in TFS 2012 is that you don't have to rely on virtualization or use System ...ter Virtual Machine Manager (SCVMM). Instead, you can use standard environments that allow you to use ...y machine, whether virtual or physical, as part of an environment. Using standard environments is the easiest way to get started with Lab Management. You only need to set up a test controller before you set up your first ...dard environment.

When you plan your pipeline, remember that there is no concrete representation of a pipeline in TFS. It is a concept. However, in the implementation presented in this book, a stage of a pipeline corresponds to a build definition. A step within a stage corresponds to a TFS workflow activity or to a shell script that is invoked by a workflow.

The following sections discuss the general approach Trey Research used to create their pipeline by using build ...plates. For a step-by-step procedure that shows you how to create the Trey Research orchestrated pipeline, ...the group of HOLs that are included under the title *Lab02-Orchestration*.

Customizing the Build Definition.

...tomizing build definitions is the easiest way to implement a continuous delivery pipeline. Each stage of the ...eline has a build definition of its own, even if the stage doesn't contain a build step. The term "build defini-...." is simply the standard name in TFS for what is otherwise known as an orchestration.

Building Once

...TFS default template always builds but a continuous delivery pipeline should build only once. For stages ...her than the commit stage, you'll need to make sure that they never create a build. The best approach is to ...the Lab Management default template for stages other than the commit stage because that template allows ...to deactivate the build step.

Propagating Versions

...ropagating versions through the pipeline doesn't happen by default. You'll need to add a mechanism in order ...one stage to trigger another. You'll also need a way to stop the pipeline if a version fails. There is some ...cussion of how to propagate a version later in this chapter. For a complete description, see the group of HOLs that are included under the title *Lab02-Orchestration*.

Creating Pipeline Instances

Build definitions don't include a way to define pipeline instances. For the sake of simplicity, the implementation presented in this guidance doesn't have a physical representation of the pipeline. It's possible to use custom work items for this purpose if you want more control or traceability but this approach is beyond the scope of this guidance

Configuring the Pipeline

...configure the pipeline, you'll need to use several editors, including Visual Studio Build Explorer and the ...ual Studio workflow editor. In general, you use Build Explorer to configure the pipeline and you use the workflow editor to configure the stages and steps. For more information, see the group of HOLs that are included under the title *Lab02-Orchestration*.

Managing Environments

To manage environments, you'll need to use Lab Management from within Microsoft Test Manager.

The New Trey Research Pipeline

After much discussion, the Trey Research team has a design for their pipeline. We're going to show you the result first, and then explain how they implemented it.

If you compare the new pipeline to the old pipeline, you'll see that the new pipeline looks very different. The team has changed its structure and added orchestration. These changes are the foundation for what they'll do in the future. Currently, not much happens right now except in the commit stage. The full implementation will occur in future iterations by automating some steps and also by adding new ones.

Here's what the orchestrated pipeline looks like.

Pipeline with Orchestration

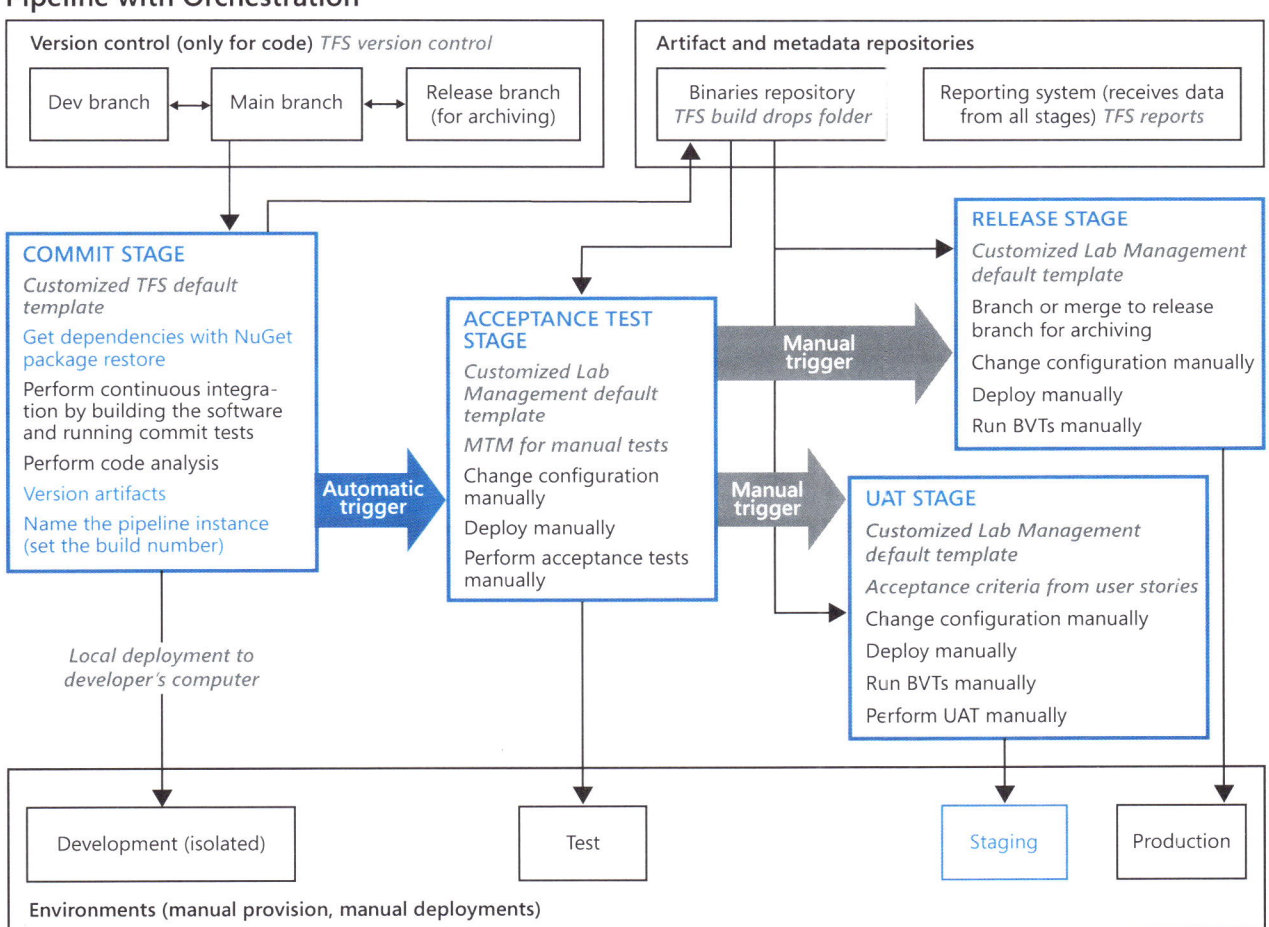

You'll notice that there's now a commit stage rather than a build stage. There are also some new steps in that stage. Along with building the binaries, the stage uses the NuGet package restore feature to retrieve dependencies. It also versions the artifacts and the pipeline instance.

If this stage succeeds, it automatically triggers an acceptance test stage. There is no longer a deployment stage. Instead, deployment is a step within every stage except the commit stage. The acceptance test stage also reflects the fact that the team is changing how they test their applications. They've begun to use Microsoft Test Manager (MTM) to plan and write their test cases.

Another difference is that the pipeline is no longer sequential. There are now two parallel stages: the release stage and the UAT stage. Both stages require manual triggers. Having parallel stages can provide faster feedback than if the stages were sequential. The assumption is that if the version has come this far through the pipeline then it will, in all likelihood, pass the remaining stages. They'll release the software without first learning their users' reactions. If later on they find that they are getting negative feedback, then they will fix it. But for now, they've gained some time. Parallel stages are the basis of techniques such as A/B testing that many continuous delivery practitioners use.

The UAT stage uses a staging environment, which is also new. Iselda finally got her new environment. Finally, the UAT stage includes another new practice for the team. They've begun to write acceptance criteria for some of their user stories.

In the next sections, we'll show you how the team uses the TFS default build template and the Lab Management default template to implement the pipeline. The implementation has two main components–orchestration and configuration. Orchestration defines the basic capabilities of a continuous delivery pipeline. Configuration defines a specific pipeline, which is the implementation shown in this guidance.

Here are Jin's thoughts on the team's efforts, in the middle of iteration 2.

Monday, August 12, 2013
The pipeline framework is done! Only the commit stage really does anything, but the rest of the stages have the placeholders we need for when we're ready to make them functional. Raymond really came through for us (much to my surprise)! Now we can concentrate on using the pipeline to release the features we're supposed to implement for this iteration. Hopefully, we can prove to our stakeholders that it's been worth it.

Here's the team's current backlog. You can see that all the pipeline tasks are done.

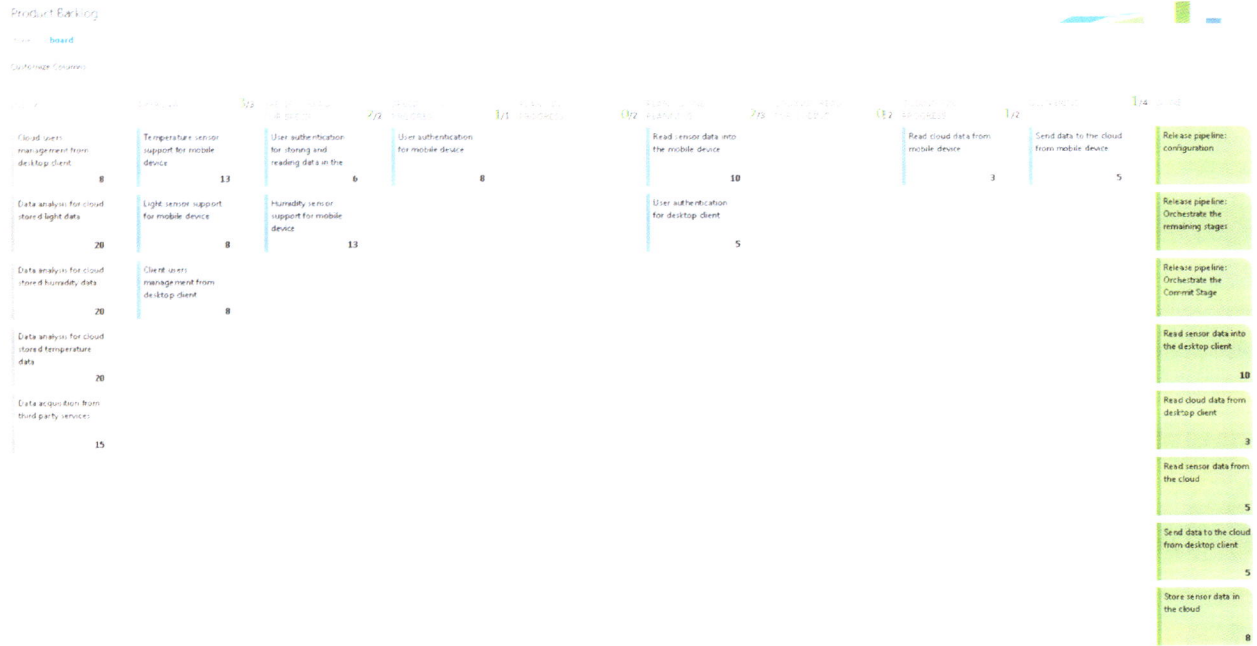

THE TREY RESEARCH IMPLEMENTATION

This section gives an overview of how Trey Research implemented the pipeline orchestration for each stage. The team customized the TFS and Lab Management default build templates as well as the build workflow (a .xaml file) to accomplish this. Again, for a step-by-step description, see the group of HOLs that are included under the title *Lab02-Orchestration*.

Orchestrating the Commit Stage

Some steps in the commit stage were implemented by customizing the TFS default build template. Other steps needed no customization because they're already included in the template. Steps that need no customization are:

- Getting dependencies. Dependencies are retrieved by using the NuGet package restore feature.
- Building the binaries.
- Running the unit tests.
- Continuous integration (in the sense of triggering the commit stage on each check-in).
- Performing code analysis.
- Copying the results to the packages or binaries repository (in the case of Trey Research, this is the configured TFS build drop folder).

The following sections discuss the steps that do require customization.

Naming the Stage and the Pipeline Instance

The name of the stage (and the pipeline instance) is generated by combining the build definition name with an automatically generated version number. The version number is generated in the standard format **Major.Minor. Build.Revision**, by using the **TFSVersion** activity from the *Community TFS Build Extensions*.

The **Major** and **Minor** parameters are provided by the team so that they can change them for major milestones or important releases. The **Build** parameter is the current date, and the **Revision** parameter is automatically incremented by TFS. This generated version number is used as the build number so that the pipeline instance can be identified. The code in version control is also labeled with the same version number in order to relate it to the binaries it generates for each pipeline instance. Here's an example.

- 0.0.0510.57 is the name of the pipeline instance. It is the same as the version number embedded in the binaries that are being built.
- 01 Commit Stage 0.0.0510.57 is the name of the commit stage for this pipeline instance.
- 02 Acceptance Test Stage 0.0.0510.57 is the name of the acceptance test stage for this pipeline instance.
- 03a Release Stage 0.0.0510.57 is the name of the release stage for this pipeline instance.
- 03b UAT Stage 0.0.0510.57 is the name of the UAT stage for this pipeline instance.

Here's the section in the build workflow where the pipeline name is calculated and assigned.

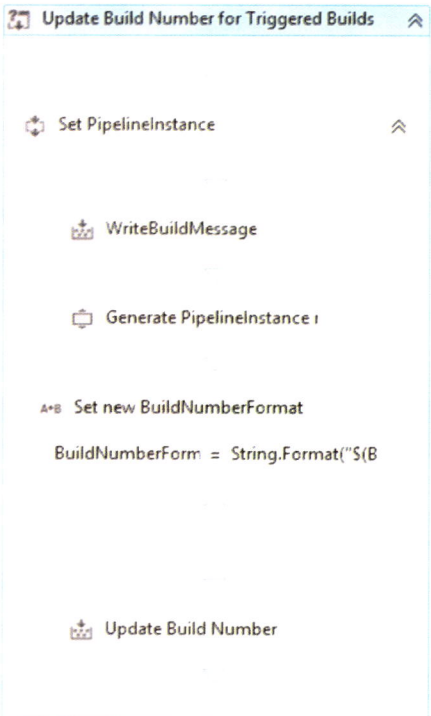

Versioning the Assemblies

The same version number that was used to name the pipeline instance is embedded in the AssemblyInfo files for the Visual Studio projects being built. By using the version number, the resulting binaries can always be related to the pipeline instance that generated them. This step is done by means of a second call to the **TFSVersion** activity.

Propagating Changes Through the Pipeline Automatically

For the last step of the build template, a list of subsequent stages is checked and the appropriate stage(s) are triggered. The list is provided by the team as a parameter for the build definition when the stage is configured. Stages are triggered by using the **QueueBuild** activity from the Community TFS Build Extensions. Two parameters are passed to the triggered stages.

- The name of the pipeline instance (the calculated version number). The name is used in the stage to set the build number. By supplying a name, you can identify stages that ran in the same pipeline instance.
- The location of the binaries repository of the pipeline instance (the drop folder). By supplying the location, the stage can find the correct binaries and not have to rebuild them from the source code.

Here's an example of how to specify and pass in the parameters. They are assigned to the **ProcessParameters** property in the **QueueBuild** workflow activity.

```vb
Visual Basic

Microsoft.TeamFoundation.Build.Workflow.WorkflowHelpers.SerializeProcessParameters(New Dictionary(Of String, Object) From {{"PipelineInstance", PipelineInstance}, {"PipelineInstanceDropLocation", BuildDetail.DropLocation}})
```

Stopping the Pipeline

If a stage fails, then the pipeline instance must stop. This step is created by using a simple conditional workflow activity, and by checking the appropriate properties.

Orchestrating the Remaining Stages

Orchestrating the remaining stages of the pipeline is partially done by customizing the Lab Management default template and also by using activities that are already in that template. These steps require no customization (at least in this iteration):

- Automatically deploying artifacts. This feature is implemented in a later iteration.
- Automatically running tests. This feature is implemented in a later iteration.
- Choosing the environment the stage uses.
- Blocking other pipeline instances from using an environment while it is being used by the current instance.

The remaining steps are implemented by customizing the Lab Management default template.

Naming the Stage

The stage is named after the pipeline instance to which it belongs. The build number is changed by using the pipeline instance name that the stage receives as a parameter. The parameter can either come from the former stage in the pipeline if the stage is automatically triggered, or provided by a team member as an argument for the build workflow if the stage is manually triggered. Here's an example of how to specify the metadata for the **PipelineInstanceforManuallyTriggeredStages** parameter.

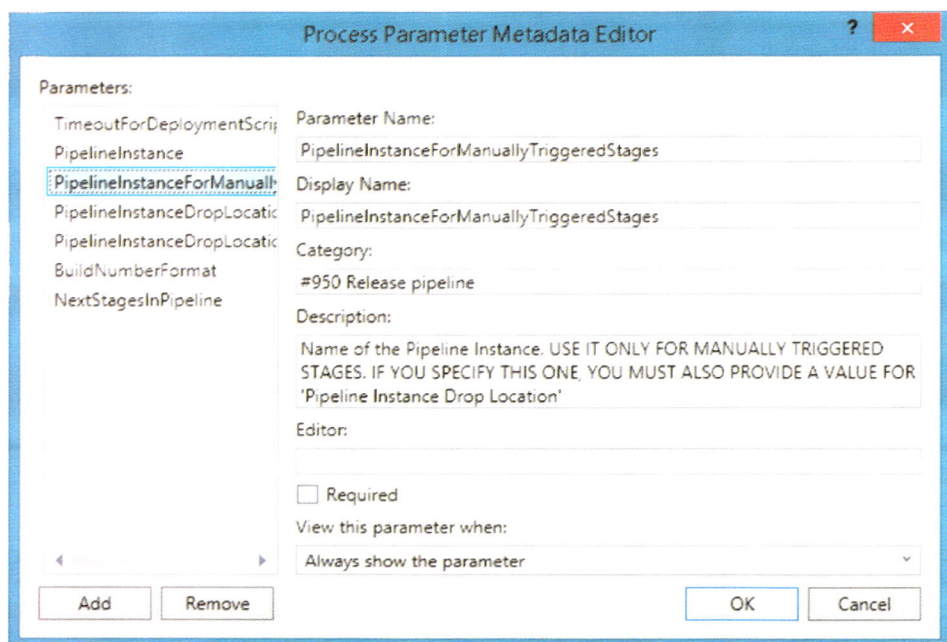

Building Only Once

To build only once, remove the portion of the workflow that builds the binaries from the template.

Retrieving the Location of the Binaries for the Pipeline Instance

The build location is changed by using the location that the stage receives as a parameter. The parameter can either come from the former stage in the pipeline if the stage is automatically triggered, or provided by a team member as an argument for the build workflow if the stage is manually triggered.

Propagating Changes Automatically

Propagating changes through the pipeline automatically is done in the same way as for the commit stage.

Stopping the Pipeline

Stopping the pipeline is done in the same way as for the commit stage.

Configuring the Pipeline

Remember that there is no concrete entity called a pipeline in TFS. For the implementation presented in this guidance, you configure the pipeline by configuring the stages. You do this by configuring the respective build definitions. Both standard parameters and the ones that were added by customizing the templates are used.

Configuring the Commit Stage

To configure the commit stage provide the following parameters to its build definition.

- The trigger mode, which is continuous integration.
- The Visual Studio solutions or projects to be built.
- The location of the binaries repository (this is known as the drop location).
- The set of unit tests to be run.
- The major version to be used for generating the pipeline instance name and for versioning the assemblies.
- The minor version to be used for generating the pipeline instance name and for versioning the assemblies.
- The list of stages to be triggered after the commit stage, if it succeeds. In this case the list has only one element—the acceptance test stage.

The following screenshot shows an example of how to configure the commit stage build definition's Process section.

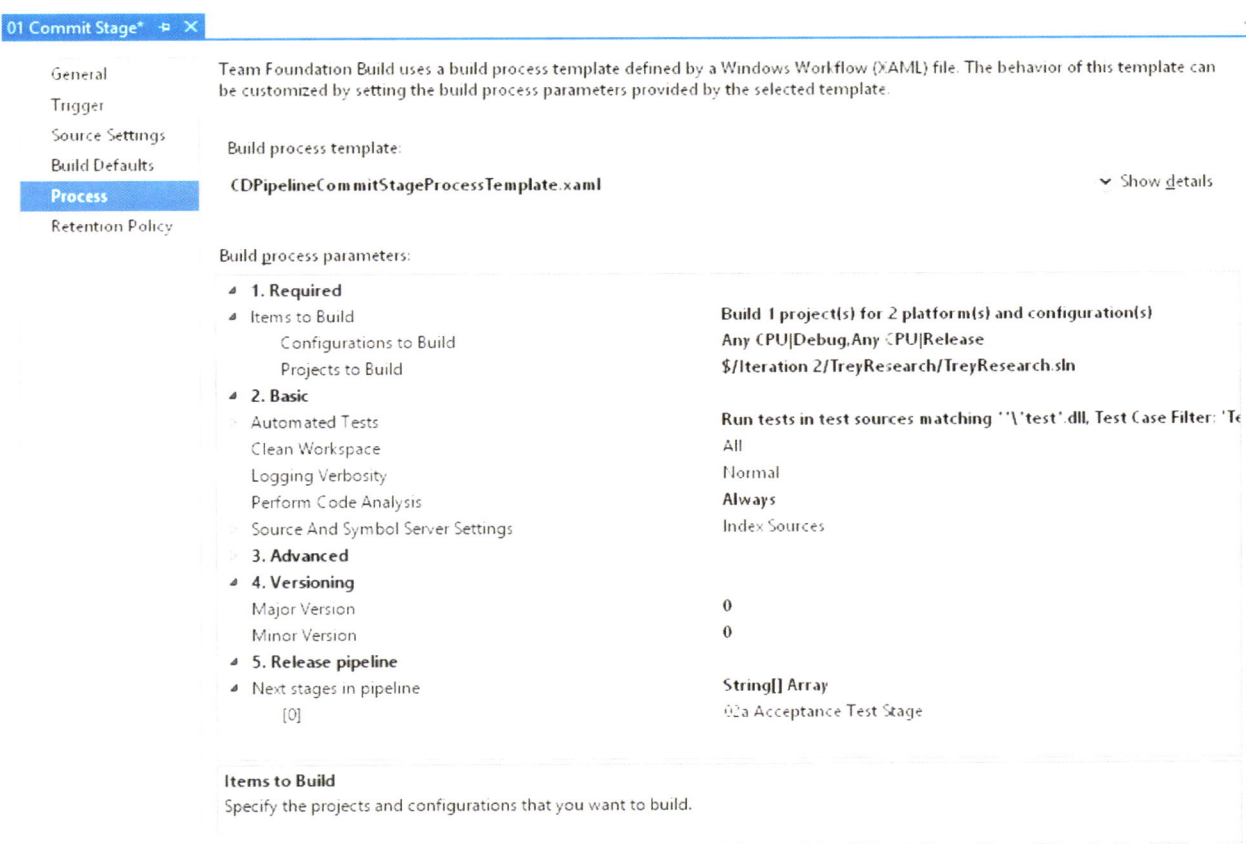

Configuring the Acceptance Test Stage

Configure the acceptance test stage by configuring its build definition. Add the list of stages to be triggered after this one, if it succeeds. In this case the list is empty because the next stages in the pipeline are manually triggered.

Because the acceptance test stage is an automatically triggered stage, the pipeline instance name and the pipeline instance drop location are passed as parameters by the commit stage.

Configuring the Release Stage

Configure the release stage by providing the following parameters to its build definition.

- The list of stages to be triggered after this one, if it succeeds. In this case, the list is empty because there are no stages after the release stage.

- Values for the pipeline instance name and the pipeline instance drop location. These parameters must be supplied because the release stage is manually triggered. The instance name can be obtained by manually copying the version number that is part of the commit stage name. The drop location is the same as the one used to configure the commit stage.

Because this is a manually triggered stage, the configuration is done by a team member just before the stage is triggered. This might occur, for example, just before releasing the software, when the team will probably want to deploy and run some tests. The following screenshot shows an example of how to configure a manually triggered stage.

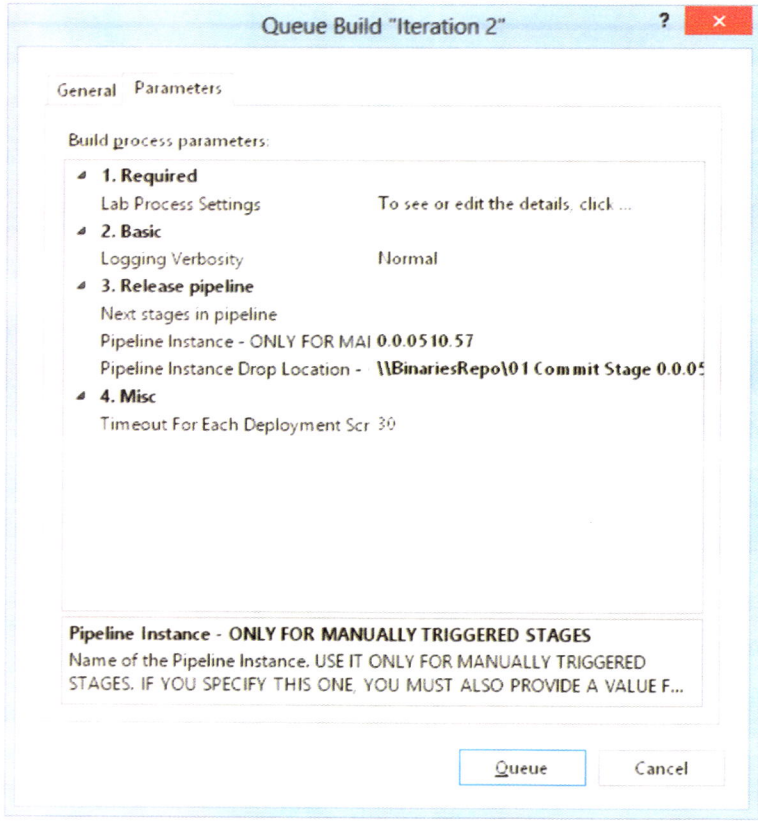

Friday, August 16, 2013
It wasn't as bad as I thought. We even managed to deliver new features for the mobile client and by "deliver" I mean coded, tested, and deployed using the new pipeline. Raymond was completely involved, worked with me every day, and the results speak for themselves. I'm really impressed with what he's done. Now I only have to get Iselda and Paulus on board for us to have a real DevOps mindset. It shouldn't be that hard because they're happy to see we're making progress.
Even though we didn't deliver all the features we planned, the stakeholders were so excited to see new features released in only in two weeks that they didn't care too much. Now we have some credibility. Next step—automation!

Configuring the UAT Stage

The UAT stage is configured the same way as the release stage.

JIN'S FINAL THOUGHTS

Here are Jin's thoughts at the end of the second iteration, when the pipeline orchestration is complete.

Here's what the team's product backlog looks like at the end of the iteration.

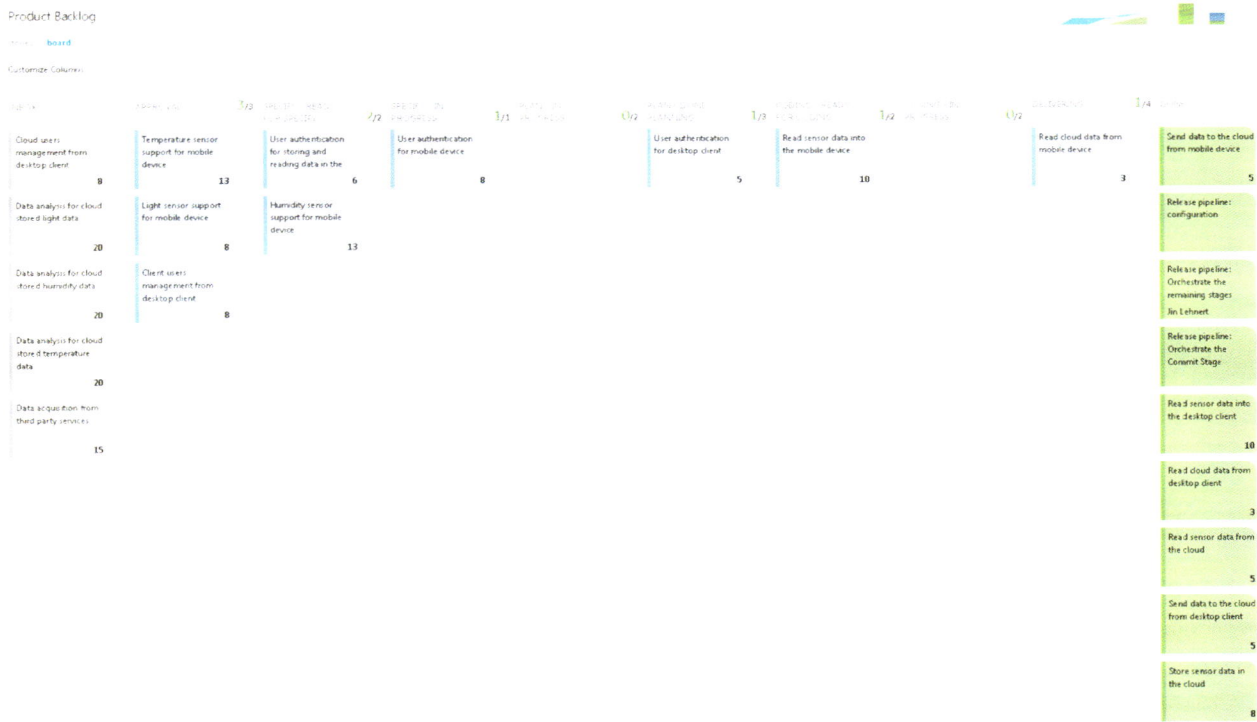

SUMMARY

In this chapter we talked about orchestration, which is the arrangement, coordination, and management of the pipeline. The goal is to take the initial steps towards creating a continuous delivery pipeline. There is some general guidance to follow such as building only once, automating as much as possible, and stopping the pipeline if a stage fails.

We also took another look at the Trey Research team, who are trying to find ways to solve their problems. They first needed a way to decide which of their many problems they should solve first. Zachary points out that there's a board meeting coming up, and if they can't show a working version of their app, they'll be shut down. Realizing that business needs dictate everything, they make a list of their most pressing problems.

Next, they need an approach that will help them fix the problems. Jin proposes a continuous delivery pipeline but gets some resistance from Iselda, who's worried about adopting new tools and the amount of works that's necessary. There's a lot of resistance from Raymond, who refuses to allow an automatic deployment to the release environment.

They finally begin to implement the pipeline and realize that it will change everything. As a result, they revise their value stream map and their Kanban boards.

Lastly, we showed the design of the new pipeline and explained how they orchestrated and configured it. The key is to use the TFS default build template and the Lab Management default template, customizing them where necessary.

WHAT'S NEXT

The Trey Research team has finished orchestrating their pipeline, but it's just the framework. Only the commit stage contains steps that are completely functional. They're still largely dependent on manual steps, which means that they still have a long list of problems. Many of them can only be solved by automating their deployments, the creation of environments, and at least some of their tests.

MORE INFORMATION

There are a number of resources listed in text throughout the book. These resources will provide additional background, bring you up to speed on various technologies, and so forth. For your convenience, there is a bibliography online that contains all the links so that these resources are just a click away. You can find the bibliography at: *http://msdn.microsoft.com/library/dn449954.aspx*.

For more information about BVTs (smoke testing), see *http://en.wikipedia.org/wiki/Smoke_testing#Software_development*.

For more information about semantic versioning, see *http://semver.org/*

For more information about using NuGet and the package restore feature, see *http://docs.nuget.org/docs/workflows/using-nuget-without-committing-packages*

The Community TFS Build Extensions are on CodePlex at *http://tfsbuildextensions.codeplex.com/*

The hands-on labs that accompany this guidance are available on the Microsoft Download Center at *http://go.microsoft.com/fwlink/p/?LinkID=317536*.

4

Automating the Release Pipeline

In the last chapter, the Trey Research team took the first steps towards improving their release pipeline and their processes. The most important step is that they've begun to talk to each other. Each team member has a viewpoint and a set of priorities. Sometimes these viewpoints clash and compromises have to be made.

To formalize their understanding of how they develop software, the team has begun to use a Kanban board, which reflects everyone's participation in the business. They've also begun to document what their product should do by using Microsoft Test Manager (MTM) to capture user stories and to develop tests.

In terms of the pipeline, they've finished the orchestration. What they have now is a framework. Only the commit stage contains steps that are completely functional. This means that they still have a very long list of problems and a lot of work left to do.

To really make progress, they need to implement some automation. Right now, all the deployments occur manually, and the environments are created manually as well. As we'll see later, manual deployments are at the root of many of Trey Research's problems. They also need to start automating some of their tests. Although the team has made progress by adopting MTM, they still run all their tests manually.

To introduce you to the topic of automation, we'll first explain the benefits it brings, (as well as some possible issues), and some general principles to follow. Again, while much of this guidance is true for any release pipeline, this guidance is tailored towards the creation of a continuous delivery pipeline. Later in the chapter, we'll show you how Trey Research automates their pipeline.

UNDERSTANDING THE BENEFITS OF AUTOMATION

Implementing automated processes can sometimes seem threatening. People may feel that their skills are undervalued or considered unnecessary, that they will lose control over the way they work, or forfeit precious resources (computers, servers, networks, software, or even people) that they have struggled to acquire. It's impossible to say that these fears are always unjustified, but automation does bring many benefits. Here are some of them.

- Automation frees people from performing repetitive, monotonous work. Not only is this work dull, but when people are bored they tend to make mistakes.
- Automation gives people more time to work on creative ways to provide value to the company.
- Automated processes are faster than their manual counterparts.
- Automation improves the quality of the entire release process because it standardizes important steps such as deployments. Well understood processes and environments result in predictable outcomes.
- Automation reduces costs, if you take a medium or long term viewpoint. Once they're in place, automated processes cost less than manual processes.

Of course, automation isn't always easy. Here are some problems people frequently encounter.

- The initial cost can be high, and it requires the dedication of the team and an investment in tools.
- The learning curve can be steep. People may not have the necessary skills or be familiar with the necessary tools.
- You may need the cooperation of people in different silos or departments.
- Automation can be very complex, depending on the scenario you want to automate. It often requires simplification. Taking a complex, fragile manual process and automating it creates a complex, fragile automated process. When you automate, always look for ways to simplify and standardize.
- If automation is done incorrectly, security can be compromised. Powerful automation tools in the wrong hands, or not used knowledgably, can cause real havoc. You should always have someone who monitors automated processes to make sure that they're functioning as they should and that they're being used correctly

In all, we feel that automation is worth it. Of course, you always have to evaluate if the return on investment resulting software quality outweigh the costs.

WHAT CAN BE AUTOMATED?

You might be surprised at the number of tasks that are candidates for automation. Here are some good examples.

Activities You May Already Be Automating

There are many activities that you've probably already automated. You may not even realize that you've done it. A prime example is building your code. Do you know anyone who, as a standard release practice, opens a command prompt, runs the compiler manually, then the assembly linker, and then copies the binaries to an output directory? (If you've just said "yes," you may want to have a serious conversation with them.)

Another example is continuous integration. More and more software development teams have automatically triggered builds that run on a build server. These automatic builds generally do more than compile the code. For example, they run unit tests, perform code analysis, and version the code. If you aren't performing continuous integration, you need to put this guidance down, start learning about continuous integration, and incorporate it to your process. You cannot have continuous delivery without continuous integration.

Activities You Should Automate

There are some activities that, if you're not automating them now, it's time to begin. These activities include:

- Deployments
- Functional tests
- Build verification tests

Deployments

Deploying an application is usually a complex process. For manual deployments, the instructions can be in multiple documents that are spread across several groups of people. These documents can easily become out of date or be incomplete. The manual deployment process must be repeated for each environment. Errors are common and costly. In short, automated deployments are a great way to improve your release process.

There are now some tools that can help you to create automated deployments. One is the ALM Rangers tool named *DevOps Deployment Workbench Express Edition*. You can learn more about it by reading Appendix 1. Another is InRelease. See Chapter 1 for a short description, or go to the *InRelease website*.

Functional Tests

Functional tests, for our purposes, are tests that ensure that the components that make up an application behave as expected. The Trey Research application has three components: the WCF service, a WPF application, and a Windows Phone 8 app. There should be automated tests in place that test all of them.

Build Verification Tests (BVT)

Build verification tests are also known as smoke tests. They ensure that the application, at some predefined level, performs as it should. If an application successfully passes the BVTs, you have a reasonable level of confidence that it will work as it should.

ACTIVITIES YOU CAN AUTOMATE

There are many activities that, although they are suitable for automation, often remain as manual tasks. Some of these often overlooked activities include performance and load testing, and what is informally called "ility testing," which covers many areas but includes, for example, scalability, extensibility, and security. Another example is automating the management of the artifact repository.

Activities That Must Remain Manual

Certain activities must remain manual. These are the activities that require a human being to do some type of validation. User acceptance tests (UAT) are a good example. These tests require that someone use the software to see if it meets some mutually-agreed upon set of requirements.

PATTERNS AND PRACTICES FOR AUTOMATED DEPLOYMENTS AND TESTS

This section discusses some patterns and practices that you can use to inform your approach to automating deployments and tests. Note that we won't be talking about testing as a subject in and of itself. There are many resources available to learn about testing. A good place to start is *Testing for Continuous Delivery with Visual Studio 2012*.

Strive for Continuous Improvement

If you don't already have automated deployments and run automated tests, then you must prepare for changes in how you work as you implement them. This guidance advocates an incremental approach to change. Examine one of your most difficult problems, find something you can do to improve it, implement it, and review your results. Apply this process to each of your most pressing issues and you will see gradual, but continuous, improvement. In this guidance, we've used Trey Research to illustrate an incremental approach to improving the pipeline, and we've also tried to demonstrate it in the hands-on labs (HOL).

Automate as Much as Is Feasible

If anything can be automated at a reasonable cost, do it. Keep manual steps to a minimum (manual steps include using an administrative graphical user interface). Ideally, get rid of them all and write scripts instead. In this guidance, we focus on automated deployments and automated tests.

Automate Deployments

Automated deployments include not only the deployment of your application, but also the automated creation of the prerequisite environments. A typical example is a web server that hosts a web application. Create a script that sets up the web server before you deploy the application to it. Automated set ups and deployments mean that you can deploy your application quickly, and as many times as you want. As a result, you can test your deployments just as you do your application.

Automate Tests

The same principles apply to tests as to deployments. Any tests that are currently done manually but that can be automated, should be. Take a look at how often you perform a particular manual test. Even if you've done the test just a few times, you know that you're wasting resources and keeping an error-prone activity in your release process. Automated testing does more than provide faster and more robust tests. It also protects you from introducing regressions and, if artifacts such as configuration files are kept in version control, lets you know if any changes to the system have broken something.

Deploy the Same Way to Every Environment

Deploy to every environment using the same process. If you use automated deployments, this means that the scripts and tools should behave the same way for all environments. Differences between environments should be externalized in configuration files.

This doesn't mean that you have a single deployment script. In fact, it's quite possible to have multiple scripts. Although it's outside the scope of this guidance, different stages of the pipeline can have different deployment requirements. For example, if you deploy to a capacity testing stage, the deployment script may prepare the agents that generate the load. This step is unnecessary for other stages. What does remain the same is that you deploy these agents to one environment the same way as you do to another environment. For example, you would deploy the agents to a test environment the same way that you would deploy them to a staging environment.

Another example of multiple deployment scripts is having one script for each component that makes up the application. As you'll see, this is true for Trey Research.

Tokenize Configurations

To tailor a deployment to a target environment, use tokens or parameters that can be provided externally to the deployment script. Candidates for tokenization include information that is dependent upon a particular deployment, the version being deployed, or the target environment. There should also be a base configuration file that contains the information that is constant for all the environments. The tokens are applied to the base configuration to create a particular deployment script.

You should also tokenize any configuration information that you need for your automated tests. For example, if your tests run against remote target URLs, extract them so that they can be provided as run-time parameters to the tests. By using tokenization, you can run the same tests across different computers or environments.

Avoid hardcoding environmental and version-specific configuration information, or any other variable data. Avoid changing this data manually to conform to another environment.

Automate the BVTs

Automated deployments should be validated by automated BVTs. These tests should give you at least a minimum level of confidence that the deployment is correct. Here are examples of what BVTs should cover.

- If you deploy a web service, test to see if the service is available.
- If you deploy changes to a database, test to see that the database is available.
- Test to see if the configuration for the target environment is correct.

Keep Everything Under Version Control

Many people keep their source code versioned, but it isn't as common to keep configuration information under version control. The configuration of the different components of an application changes during the product's lifecycle, and it's important to be able to recreate any of these versions, if necessary. You should also save deployment scripts and any artifacts that are a part of an automated deployment in your version control system. The same is true of configuration files that you use with your automated tests.

Any tools that you use for deployments are also good candidates for version control. By following this practice, the right versions of the tools will be available if you ever need to recreate a dep oyment for any specific version of the software. Keeping tools under version control is particularly relevant for applications that are only sporadically maintained after they're released.

You may not want to store all the tools in TFS version control, especially if they take up several gigabytes. The goal is that the information is stored somewhere other than in people's heads or in a document. For example, you can prepare a virtual machine (VM) that is configured with all the necessary tools, and store it in the library of a virtualization system such as System Center Virtual Machine Manager (SCVMM). You may want to version the VM as the tools are updated.

Use One-Click Deployments

You should be able to run fully automated deployments with little effort. If you need to run a series of scripts manually, then your deployments aren't fully automated. The ideal situation is that the only manual action required is when someone starts the deployment process.

Build Once

Although the best practice of building your code once was already discussed in the last chapter, we're bringing it up again in the context of the deployment scripts. None of them should rebuild the application from the source code. Building once ensures binary integrity. This means that the code that is released into production is guaranteed to be the code that was tested and verified throughout all the pipeline stages.

Although building once seems straightforward to implement, it's easy to mistakenly build multiple times with TFS Build. Typically, people use the TFS default build template to create a build definition that triggers the deployment. This template is designed to rebuild the code just before the deployment occurs. The same is true for configuration file transformations. Generally, if you want to transform a configuration file by using the standard procedure in Visual Studio or TFS Build, you have to rebuild the application by using a different build configuration. The issue is that you don't want a different build configuration for each environment. You want to use the same build configuration (typically, the Release configuration) that was created in the commit stage, with different application configuration files.

Another way to ensure a single build is for the commit stage to store the binaries it builds in a repository. All the deployment scripts should retrieve the files they need from that repository, and never from the previous environment in the pipeline.

Choose a Suitable Deployment Orchestration Model

As you form a plan for automating your deployments, consider the deployment orchestration model, which defines how the deployment script executes. Deployment-level orchestration is an example of orchestration at the step level of the pipeline. Among the decisions to be made when you begin to automate deployments is what you'll use for the deployment orchestration model, which defines the way the deployment scripts execute. There are three options to choose from.

Use a Tool that is Native to Your Platform

The first option is to use a deployment management and orchestration tool that works natively with your platform. For the Windows platform, several tools are available. One possibility is *Active Directory Group Policy (GPO)*. Another possibility is to use a combination of *System Center Service Manager and System Center Orchestrator*. Both of these options rely on *Windows Installer*, which is the native packaging tool for the platform.

Because it is the native deployment technology, Windows Installer is a useful deployment tool even if GPO and System Center aren't available. You can write deployment scripts that use the command line tool, msiexec.exe, to automatically deploy the .msi file that contains the Windows Installer package.

Using tools that are native to your platform is the most direct approach for orchestrating deployments. These tools already have the capabilities to perform many important functions. They can install dependencies, deal with versioning and upgrades, schedule deployments, determine the states of the target machines, and validate the security of scripts. Using native tools means that you use the same tools to manage deployments as you use to manage the infrastructure, which simplifies both tasks. Also, operations people are usually familiar with these tools.

Again, as an example, Windows Installer supports all these features and many others. It's also not limited to desktop applications. You can, for instance, prepare an MSI installer that deploys a website simply by configuring the necessary steps with a utility such as the WiX toolset. For more information, go to the *Wix Toolset* website.

Use Deployment Agents that Are Provided by the Release Management System

The second option is to use deployment agents that are provided by the release management system. A deployment agent is a lightweight service that's automatically installed on the target machine and that runs the steps needed to deploy locally to that computer. These steps are typically contained in a script that is written in a technology that is natively supported by the target operating system (possibilities for Windows include batch scripts and PowerShell).

With TFS, the deployment agent is provided by Lab Management. The agent is automatically installed on the target computers when they are added to a Lab Management environment. This approach is not as powerful as the first option, but still has many advantages. The scripts are much simpler than those required by the third option, remote scripting, because they run locally. Also, the release management system (which, in this guidance, is TFS Build and Lab Management) performs many tasks, such as orchestrating the deployment, logging, and managing the target environments.

This approach is also secure. The environments are locked down and only the service account that runs the scripts has permission to change them.

Use Scripts that Connect Remotely to the Target Machines

The third option is to perform the deployment by using scripts that connect remotely to the target machines. For the Windows platforms, there are several options such as *PsExec* and PowerShell. This is the least powerful option. You will have to do all the orchestration work yourself, and security can be harder to enforce.

For more information about these deployment orchestration models, you can refer to the "Deployment Scripting" section in Chapter 6 of Humble and Farley's book, *Continuous Delivery*.

Choose a Suitable Testing Orchestration Model

You will also need to choose an orchestration model for your automated tests. There are two options.

Run the Tests Locally Using a Testing Agent

The first option is to run the tests locally, on the target machine, by using a testing agent that is provided by your release management system. An example is the Lab Management test agent. The advantages of this orchestration model are similar to those for deployment. The test code is simpler and needs less configuration, a great deal of the work is done by the release management system, and the security is good because the tests don't run across the network or outside the context of the target computer.

Run the Tests Remotely

The second option is to run the tests remotely by having them connect to the target environments. You can use this approach if you run tests (for example, integration tests or UI tests) from a build agent, in the context of a build definition that is based on the default TFS template. You'll have more configuration, orchestration, and management work to do in order to make the tests suitable for different environments, and security will be weaker than if you use a test agent. However, there may be some cases where this is the only available option.

Follow a Consistent Deployment Process

Whenever possible, follow the same deployment process for all of the components in your system. It's much easier to set up and maintain a single, standardized approach than multiple approaches that have each been tweaked to work with a particular environment. By stressing a uniform process, you may also find that you can reuse some of the scripts and artifacts for several of your system components. Using the commit stage to prepare the artifacts and then running a script in the subsequent stages is one way to follow a consistent deployment process.

Use the Commit Stage to Prepare the Artifacts

The commit stage should prepare any artifacts you'll need to deploy to the other stages of the pipeline. Typically, these artifacts include the deployment packages and the configuration files that are tailored to the target environments. If you want your commit stage to complete quickly in order to get fast feedback about the build, you might want to split it into two stages. The first stage can run the tests that let you know if the build succeeded. The second stage can prepare the configuration files and the packages. The second stage should be automatically triggered when the first stage completes successfully.

Run a Script in the Subsequent Stages

After the commit stage, the subsequent stages should run deployment scripts that they retrieve from version control. These script should retrieve the required packages and configuration files from the binaries repository.

Leave the Environment and Data in a Known State

Automated deployments should leave target environments in a known, consistent state, so that they can be used immediately without having to perform further steps to clean them up. This is also true of any databases or data repositories that the application uses.

In terms of the target environments, the easiest approach is to always perform complete deployments of every component. Incremental deployments, where you deploy only the artifacts that have changed, are much more complex because you have to keep track of what changes and what doesn't, and test how the new components work with the old ones.

For databases, leave the target database in a state that is compatible with existing and new data, and that has a schema that is compatible with the version of each component. Although the subject of managing changes to databases during deployments is outside the scope of this guidance, there are several approaches that you can investigate. Here are three of them.

- Schema and data synchronization, which you can implement with tools such as *SQL Server Data Tools*.
- Delta scripts, which you can write with *DBDeploy.NET*.
- Database migrations, which you can implement with object-relational mapping tools such as *Entity Framework Migrations*.
- *Redgate Software* has a variety of tools for SQL Server development and database management.

In terms of automated testing, maintaining consistent environments and test data suites is especially relevant because the tests depend on having a known initial state. Again, the preparation and management of useful test data suites is outside the scope of this guidance. To get started, read Chapter 12 of *Continuous Delivery* by Jez Humble and David Farley.

Have a Rollback Mechanism Available

While useful for any environment, rollback mechanisms are particularly important for production environments. A rollback mechanism allows you to return the target machines to an operational state if anything goes wrong during a deployment. With a rollback mechanism in place, you can ensure that the application is still available while the problem that caused the current deployment to fail is investigated. Of course, an automated rollback mechanism should be your goal.

The easiest way to perform a rollback is to redeploy the version that was running just before the failed deployment. If the automated deployment scripts are written to work with any version of the application (See "Tokenize Configurations" earlier in this chapter), you can use these scripts as the rollback mechanism. You only need to provide the scripts with the version you want to deploy.

Lock Down the Environments

Once you have automated the deployments, there is no reason to allow users to change environments manually. You should only modify the target machines by running the deployment scripts. As a rule of thumb, only a few user accounts should have enough privileges to run these scripts. This restriction ensures that everyone uses the automated procedures and it stops people from manually trying to change the environments, or using procedures other than the correct scripts.

If you use Lab Management build definitions or remote deployment, the user accounts that can run deployments should be limited to:

- The service accounts that run the deployment agents.
- Users with administrative privileges.

This way, deployments or any kind of environmental change are only possible if done by the pipeline, or if invoked purposely by users with the correct privileges. An example of when this second case might occur is if you need to perform a rollback.

Make Deployment Scripts Granular

Although we've stressed the importance of uniform procedures, this doesn't mean that a single script that deploys an entire system is the best way to go. Instead, make your deployment scripts granular. By this we mean that each deployment script should focus on a particular component.

In terms of deployments, a component is a set of artifacts (binaries, configuration files, and other supporting files), that can be deployed together, without interrupting other areas of the application. For .NET Framework projects, the organization of these components ultimately depends on the Visual Studio (MSBuild) projects that make up your solution.

Having different deployment scripts for different components lets you deploy changes to individual components without having to discontinue service or affect other components in the system. The deployment process itself will be shorter, and there will be fewer chances for errors. Furthermore, if any problems occur, they will be easier to solve.

It may seem that writing deployment scripts that act on the component level contradicts the advice about avoiding incremental deployments given earlier in "Leave the Environment and Data in a Known State." To distinguish between incremental and non-incremental deployments, evaluate a deployment at the component level.

For example, you can incrementally deploy a single component by only deploying the binaries that have changed. This is often more complex than the alternative, which is to deploy the entire component, including all its binaries and supporting files. At the application level, this second approach may seem like an incremental deployment, but at the component level, it isn't.

Of course, as with any guidance, you must take your own situation into account and make a decision that balances the complexities of the proposed deployment against other factors, such as how much the deployment will cost in terms of time and resources.

Adopt a DevOps Mindset

One of the key principles of DevOps is that there must be collaboration between development and operations teams. Cooperation between these two groups can definitely make it easier to automate deployments. If your goal is continuous delivery, you need the people who write the deployment scripts to collaborate with the people who manage the environments and run the scripts. Poor communication can cause many problems that can make the automation process frustrating and error prone.

When you begin to plan how to automate deployments, make sure that there are people from operations working with people from development. The same principle is true of testing. When you begin to move toward automated testing, make sure that people from test work together with people from development to create the process.

Begin the Process Early and Evolve It

The longer you wait to automate deployments and tests the harder it will be because scenarios get more complex as time goes on. If you can start early, preferably from the beginning of the project, you can begin simply and evolve the processes over time, incrementally, just as you do your software. Furthermore, you won't be able to tell if your software is doing what it should until it's been deployed and tested. The best way to discover if your software is working correctly is to automate your deployments and tests so that you get feedback as quickly as possible. Finally, it can take time to learn the tools and techniques for automation. Trying to cram it all in at the end of a project is guaranteed to cause trouble.

Choose the Right Tests for Each Stage

Not all the automated tests you write should run in every stage. The purpose of certain stages is to run specific types of tests and they should run only after other, more lightweight validations.

For example, it doesn't make sense to run load tests in the commit stage. They would slow the stage down and they would be premature because load tests should run only after the software is functionally validated. Another obvious example is that unit tests should only run in the commit stage rather than, for instance, in the acceptance test stage.

Generally, specific types of test only run once in the pipeline, as a step in a particular stage. The exception is BVTs. They validate deployments so they should run in each stage where automated deployments occur.

TREY RESEARCH

Now let's take a look at how Trey Research is implementing these patterns and practices. When we left them, they'd finished orchestrating their pipeline, but what they had was a framework that still didn't provide a lot of functionality. Consequently, the long list of problems they had at the beginning doesn't look that much shorter to them. Here are some of them.

Issue	Cause	Solution
They never know if they have the correct version. What they think is a current version isn't, and they find bugs they've already fixed, or missing features they've already implemented.	The deployed versions are out of date.	Associate deployments to specific changes. Automate the deployments and trigger them either as a step in the pipeline, or by a manual command.
They have to perform acceptance tests again and again to prevent regression errors. This is a large, ever increasing amount of work, and is both slow and error prone. Consequently, they don't test as thoroughly as they should.	All the test cases are performed manually.	Automate the acceptance tests and run them in the corresponding stage of the pipeline.
The deployment process is slow and error prone.	There is no standard deployment process. One deployment is different from another deployment. All deployments are manual.	Automate deployments to all environments.
They don't know how to deploy to different environments.	They change the application configuration manually for each environment. This occurs every time there is a deployment.	Modify the pipeline so that it changes the configuration files to suit particular environments.
Existing environments are vulnerable to uncontrolled changes.	The environments aren't locked down. Too many people have permissions to access them and make changes.	Lock down the environments so that changes occur only in the context of automated scripts that run under special user accounts or by authorized team members

There are also other pressures coming into play. The board wants the Trey Research application to have some new features. They're impatient because the work to orchestrate the pipeline took a fair amount of time. Another issue is that everyone on the team has their own priorities.

I'm still not sold on all of this. I don't think we're paying enough attention to security, and there is no way I'm letting some test agent run in the release environment.

I need to build new features. I can't spend all my time with Jin, working on the pipeline. No one's paying me for that.

I need to be working on tests. I like MTM and I think we do need to stop testing manually, but it's a lot of work.

Everyone wants something different. Our resources are limited, and if we don't add new features, the competition will win. On the other hand, I know that our release process is a mess but fixing it isn't trivial. How do I keep everything in balance?

In fact, Zachary is feeling overwhelmed.

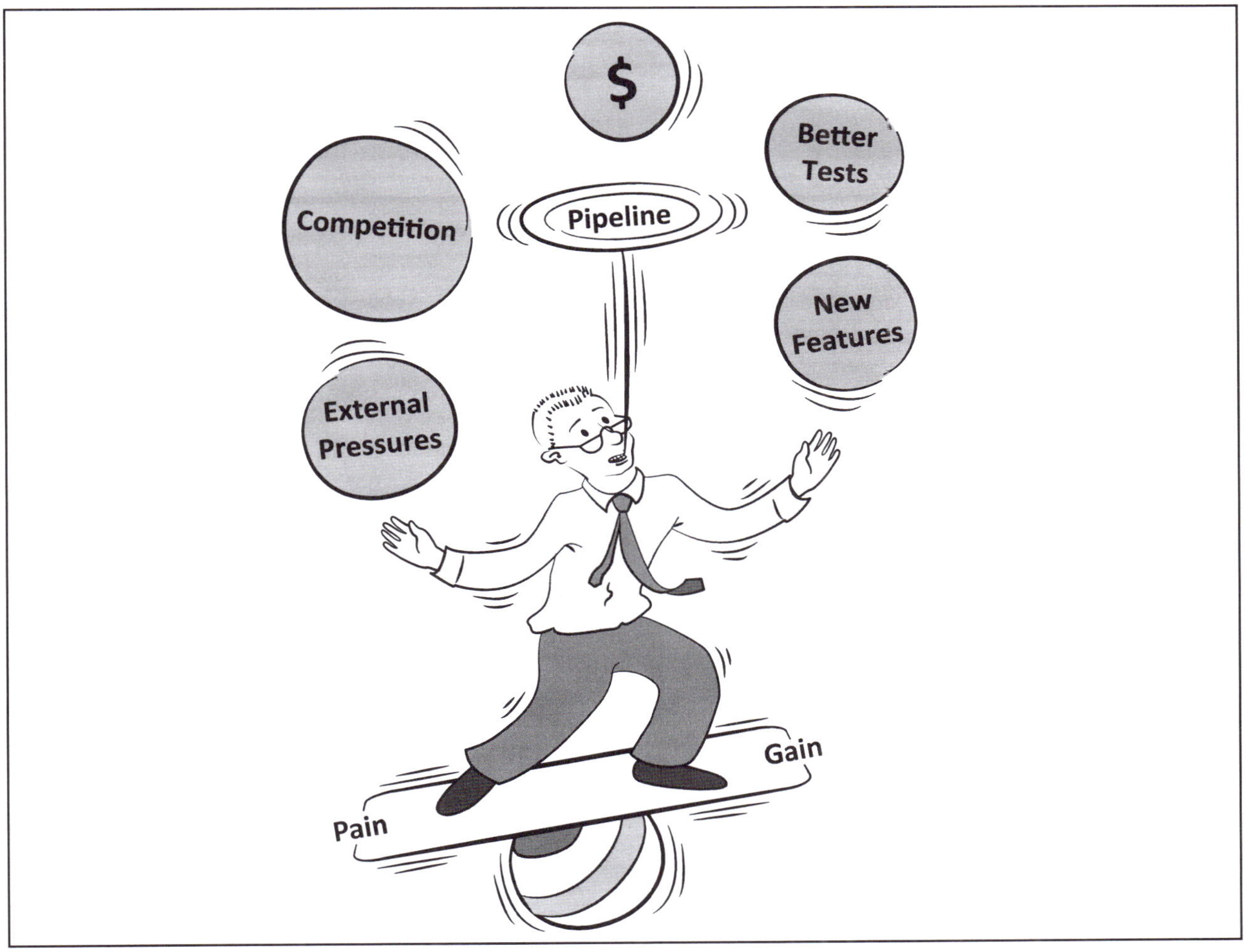

After much debate and drawing on whiteboards, the team decides to focus their pipeline efforts on automating deployments and tests. Although it will be a big push, once they're done they'll have a continuous delivery pipeline that should solve their major problems. Here's the final result of all their hard work.

Pipeline with Automation

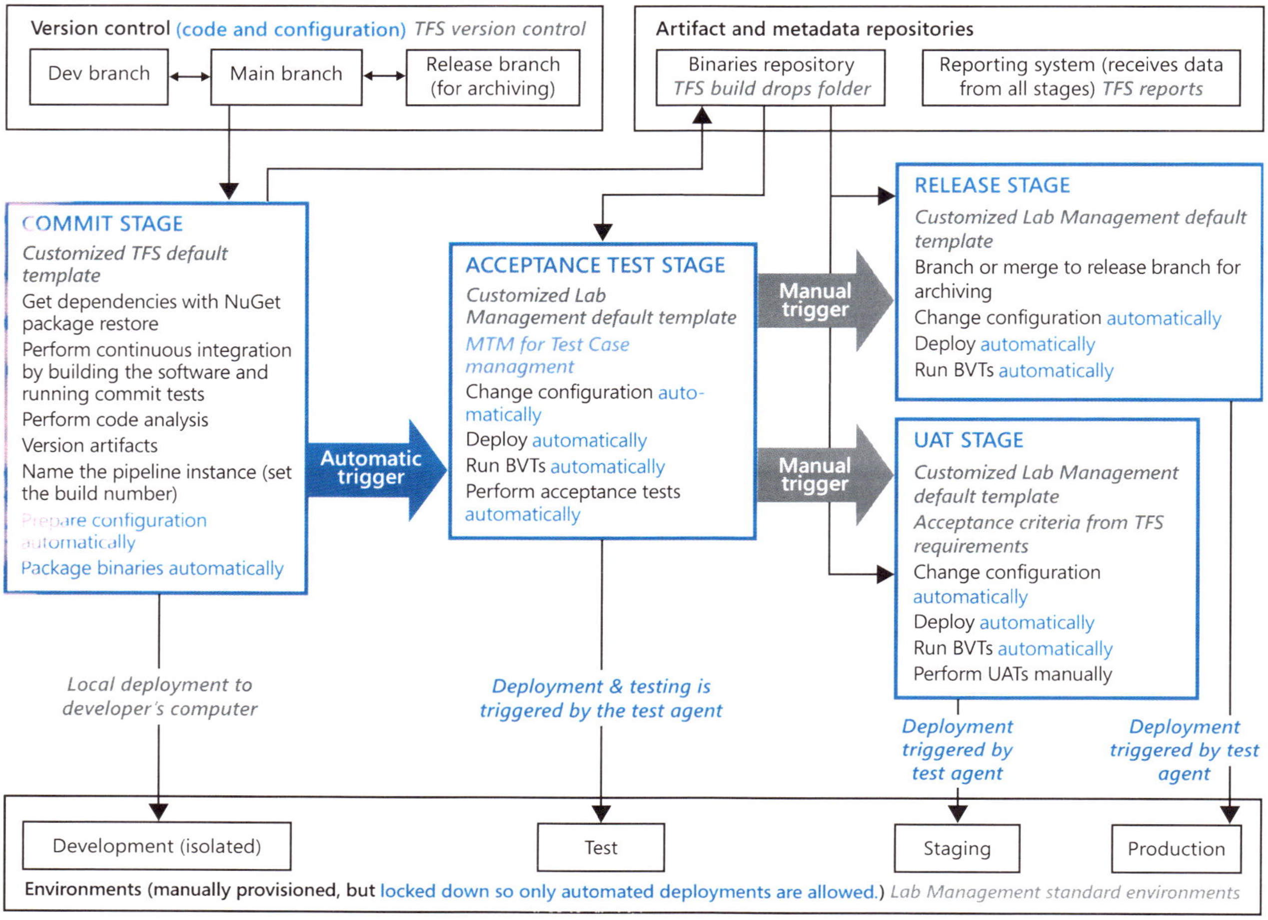

Version control (code and configuration) *TFS version control*

| Dev branch | Main branch | Release branch (for archiving) |

Artifact and metadata repositories

| Binaries repository *TFS build drops folder* | Reporting system (receives data from all stages) *TFS reports* |

COMMIT STAGE

Customized TFS default template

Get dependencies with NuGet package restore

Perform continuous integration by building the software and running commit tests

Perform code analysis

Version artifacts

Name the pipeline instance (set the build number)

Prepare configuration automatically

Package binaries automatically

Automatic trigger

ACCEPTANCE TEST STAGE

Customized Lab Management default template

MTM for Test Case managment

Change configuration automatically

Deploy automatically

Run BVTs automatically

Perform acceptance tests automatically

Manual trigger

Manual trigger

RELEASE STAGE

Customized Lab Management default template

Branch or merge to release branch for archiving

Change configuration automatically

Deploy automatically

Run BVTs automatically

UAT STAGE

Customized Lab Management default template

Acceptance criteria from TFS requirements

Change configuration automatically

Deploy automatically

Run BVTs automatically

Perform UATs manually

Local deployment to developer's computer

Deployment & testing is triggered by the test agent

Deployment triggered by test agent

Deployment triggered by test agent

| Development (isolated) | Test | Staging | Production |

Environments (manually provisioned, but locked down so only automated deployments are allowed.) *Lab Management standard environments*

The rest of this chapter explains what the Trey Research team did. At a high level, we can summarize their efforts by saying that, for automated deployments, they wrote scripts that use Lab Management agents to run locally across all the environments. They also configured the pipeline to run these scripts and to provide the correct parameters.

Here's what Jin thinks as he looks at the beginning of a new iteration.

Monday, August 19, 2013

It seems like Zachary is feeling more confident. He asked us to include two new features in the backlog. He even seems to understand that we can't start working on them if we're already busy working on other work items. We've also added a bug that a user found in the new UAT environment to the backlog. Of course, we also have all the work to automate the pipeline. We increased our WIP limit of the Ready for Coding column. Let's hope we can handle it.

Here's the product backlog for this iteration. You can see that there are many tasks associated with automating the pipeline, as well work on the Trey Research application.

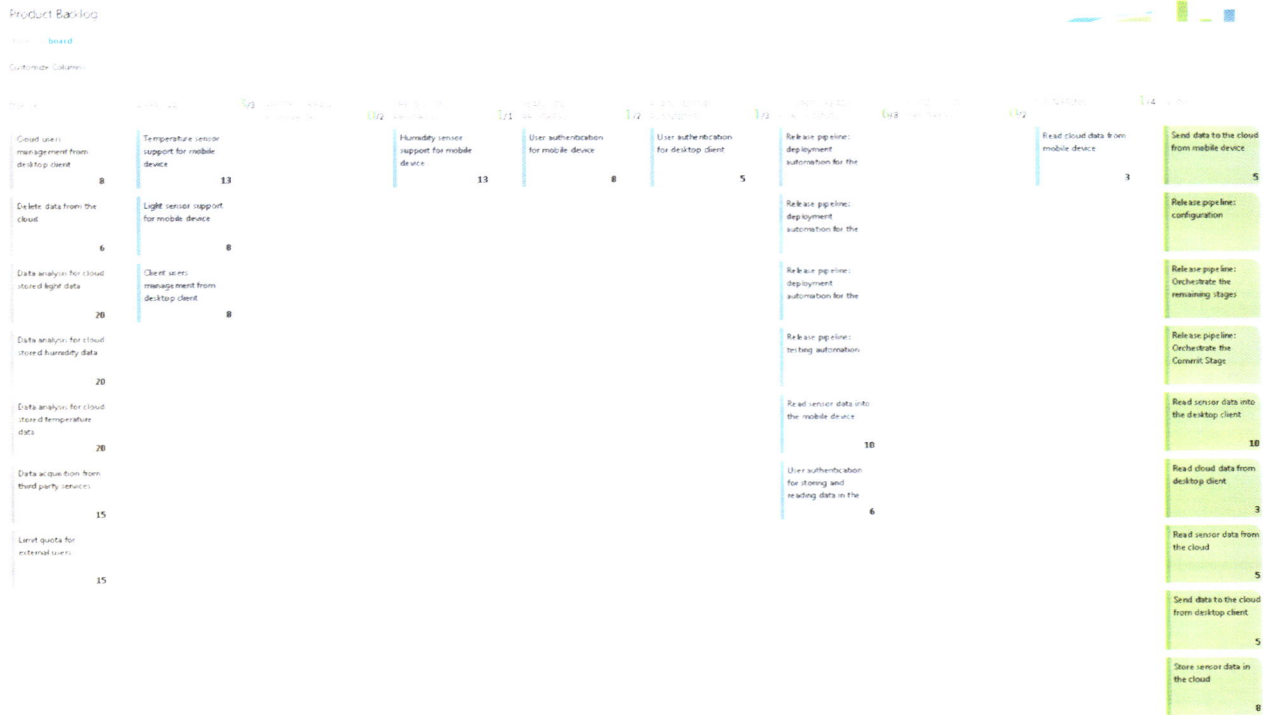

Here's what else Jin has to say.

Monday, August 19, 2013

In the first part of this iteration, Paulus (who's the only person who knows how the components of the application work) will team up with Raymond to automate their deployment. I'll help Iselda set up the automated tests. After that, we'll all concentrate on building some new features so the stakeholders are happy, and so that the pipeline automation is tested with a realistic scenario.

Here's the backlog for this iteration. You can see who's responsible for each task.

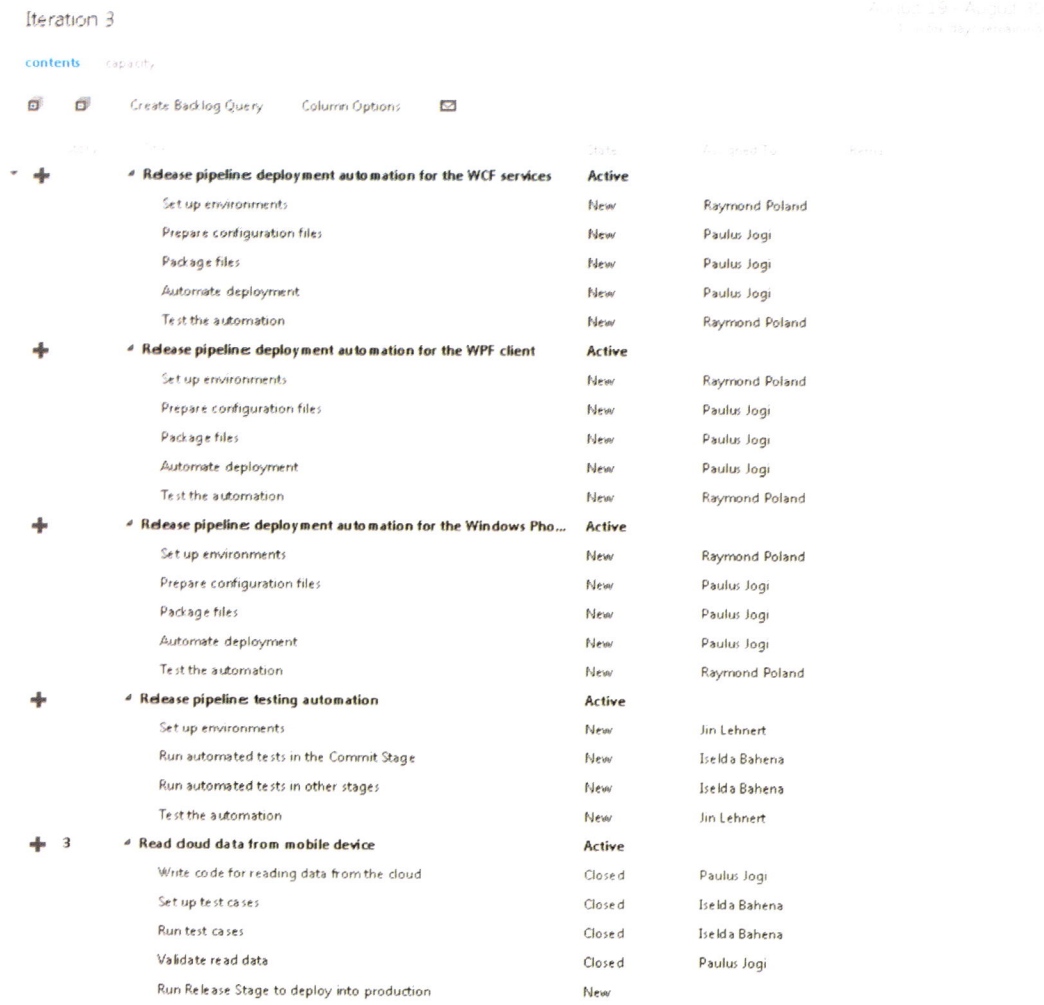

How is Trey Research Automating Deployments

Adding automation is the major goal this iteration for the pipeline. This section discusses the key points towards accomplishing that goal. For a step-by-step description of how to automate the pipeline, see the group of labs included under the title *Lab03 – Automation*. If you look through these lab, you'll see that the team wrote three different deployment scripts. There is one script for each component of the application. This means that there is a script for the WCF web service, one for the Windows Phone 8 client, and one for the WPF application. Currently, Trey Research doesn't need different scripts for the different stages because, in terms of deployments, all the stages in the pipeline have the same requirements.

They still need to write deployment scripts that set up the web server, the Windows Phone 8 emulator, and the Windows computer for the Windows Presentation Foundation (WPF) client. This is something that they plan to do in the future.

How is Trey Research Deploying the Same Way to Every Environment

Trey Research has made sure that the scripts deploy the same way to all three environments. The deployment agent retrieves a script directly from the drop location, where it is placed by the specific instance of the pipeline. They use the **$(BuildLocation)** built-in variable to compose the path. The first parameter is the drop location. The second parameter is the name of the target environment.

The following code shows how to deploy the WCF web service to the test environment.

CMD

```
"$(BuildLocation)\Release\Deployment\WcfService\DeployWcfService.cmd" "$(BuildLocation)" Testing C:\
TreyResearchDeployment
```

The following code shows how to deploy the WCF web service to the production environment.

CMD

```
"$(BuildLocation)\Release\Deployment\WcfService\DeployWcfService.cmd" "$(BuildLocation)" Production C:\
TreyResearchDeployment
```

How Does Trey Research Use Automated BVTs

Trey Research has created a set of automated BVTs that ensure that the deployments are correct. All the stages that perform automated deployments are configured to run them.

In the future, as an improvement, the team plans to have different BVT test suites that cover specific deployments. If they deploy a specific component, they'll be able to run the BVTs that verify the deployment of that component.

How Does Trey Research Tokenize Configuration

The team identified all the configuration information that was specific to a particular environment and extracted it to specific tokenization (or parameter) files. These files only contain the information that changes from one deployment to another. Trey Research uses configuration transformations to prepare the final configuration files that are used during the deployment.

There is a base configuration file for each group of environment-specific parameter files. The transformation is applied to that base file. For example, this code defines the way the value of the endpoint setting is transformed.

XML

```xml
<endpoint address="http://webServerAddress:portNumber/SensorReadingService.svc"
          name="BasicHttpBinding_ISensorReadingService"
          xdt:Locator="Match(name)"
          xdt:Transform="SetAttributes(address)">
</endpoint>
```

There is a base configuration file for each group of environment-specific parameter files. The transformation is applied to that base file, so only the portions of the file that must change are transformed.

For now, Trey Research tests are run locally by the Lab Management test agent. This means that the team doesn't need to extract any test configuration information and put it into separate files. Tokenizing test configuration information may be something they'll do in the future.

How Does Trey Research Keep Everything under Version Control

Configuration files are part of Visual Studio projects, so they are under version control. Deployment scripts are also part of their respective Visual Studio projects, so they too are version controlled. TFS keeps build scripts (workflow templates) under version control by default. TFS does not provide a direct way to version build definitions. As a future improvement, the team plans to make a copy of each build definition before it changes.

Right now, the team doesn't keep tools under version control. When they must update a tool, they'll decide on the best way to keep the old version available.

How Does Trey Research Provide One-Click Deployments

The entry point for all the deployments is a single deployment script. Once a script runs, the component is deployed without requiring any more manual actions.

If necessary, the script can be run from the command line on the target machine. However, Trey Research usually runs it as a step inside a stage of the pipeline. If the stage is automatically triggered, the script begins with no manual intervention. If it is one of the manually triggered stages, such as the UAT stage, then the script is triggered when the stage is triggered.

How Does Trey Research Ensure That They Build Only Once

Trey Research deactivated the build step in all the pipeline stages but the commit stage. They also added a step in the commit stage that transforms the configuration files explicitly. This transformation is instead of changing the build configuration and rebuilding the application each time they need to transform the application configuration files. The following code is an example of how to explicitly perform the transformation inside the MSBuild project file of a component.

XML

```
<TransformXml Source="App.config" Transform="@(TransformationFiles)" Destination="$(OutDir)\ConfigFiles\
WpfClient\%(TransformationFiles.Identity)" />
```

Finally, all the deployment scripts obtain the artifacts from the binaries repository (the drop folder for the pipeline instance), instead of getting them from the previous environment in the chain.

What Is the Trey Research Deployment Orchestration Model

Trey Research uses a hybrid approach towards deployment orchestration. It's based on the deployment agent model but relies on the platform's packaging tools. They use Lab Management, so all the deployments are run locally on the target computers by the Lab Management deployment agent. The agent is a Windows service that runs the deployment script. The script itself is obtained from the binaries repository. The packaging technology depends on the platform. For the WCF web service, they use MSDeploy packages that are supported by IIS. For the Windows Phone 8 app, they use XAP packages. For WPF, they use native MSI installers.

What Is the Trey Research Testing Orchestration Model

Trey Research's testing orchestration model is to use the build agent for the commit stage's unit tests. The subsequent stages use the Lab Management-based build definitions. The test agent that TFS provides runs the tests locally.

Lab Management only allows one machine role to run tests inside an environment. For Trey Research, this is the client role. If Trey Research wants to run tests directly against the web service, they'll have to do it remotely.

How Does Trey Research Follow a Consistent Deployment Process

Trey Research follows the same deployment process for the WCF web service, the WPF application and the Windows Phone 8 app. Here's a summary of the process.

1. The commit stage packages the files to be deployed and prepares the configuration files.

    a. The configuration files are prepared by using a base template for each component. The base template is transformed by using the **MSBuild TransformXml** task. The base template is transformed to include the environment-specific parameters and data, which results in a different configuration file for each environment.

    b. The files to be deployed are packaged using the standard tool for each technology: MSDeploy zip packages for the WCF web service, XAP packages for Windows Phone 8, and MSI Windows Installers for WPF.

    c. The packages and configuration files are copied to the binaries repository. The remaining pipeline stages retrieve them from that location.

2. The subsequent stages run a deployment script that uses the prepared packages and configuration files to deploy to the target environment. The script is run locally on the target machine by the Lab Management deployment agent.

How Does Trey Research Leave the Environment and Data in a Known State

The WCF web service, the Windows Phone 8 app, and the WPF application are versioned and packaged in their entirety. Each of them is deployed as a complete package. There are no deployments that use subsets of a component or specific files.

Currently, the Trey Research application doesn't use a database, so the team doesn't yet know which approach they'll adopt to leave a database in a known state. They also don't currently run tests that require specific suites of test data.

What Rollback Mechanism Does Trey Research Use

All the deployment scripts are tokenized and they receive the version to be deployed as a parameter. The following code shows how the deployment script for the WCF web service is called in order to deploy the service to the test environment. The **$(BuildLocation)** parameter contains the version to be deployed.

```
CMD
"$(BuildLocation)\Release\Deployment\WcfService\DeployWcfService.cmd" "$(BuildLocation)" Testing C:\
TreyResearchDeployment
An example value of $(BuildLocation) for this invocation would be "\\<Path to the binaries repository> \01
Commit Stage\01 Commit Stage 0.0.0605.781".
```

If the team needs to perform a rollback, they can run the deployment script and point it at the version that was running before the failed attempt.

How Does Trey Research Lock Down Its Environments

Trey Research locked down its environments by ensuring that only the account used to set up the Lab Management environment can change them. In addition, the operations manager, Raymond, also has administrative permissions, in case a manual change or a rollback is required.

How Does Trey Research Make Its Deployment Scripts Granular

Trey Research uses the Visual Studio (MSBuild) project that's available for each of the components that make up the Trey Research Visual Studio solution. Each of these components has a different deployment script, which is included in the appropriate MSBuild project. The following screenshot shows the location of the deployment script for the Trey Research application's WCF service.

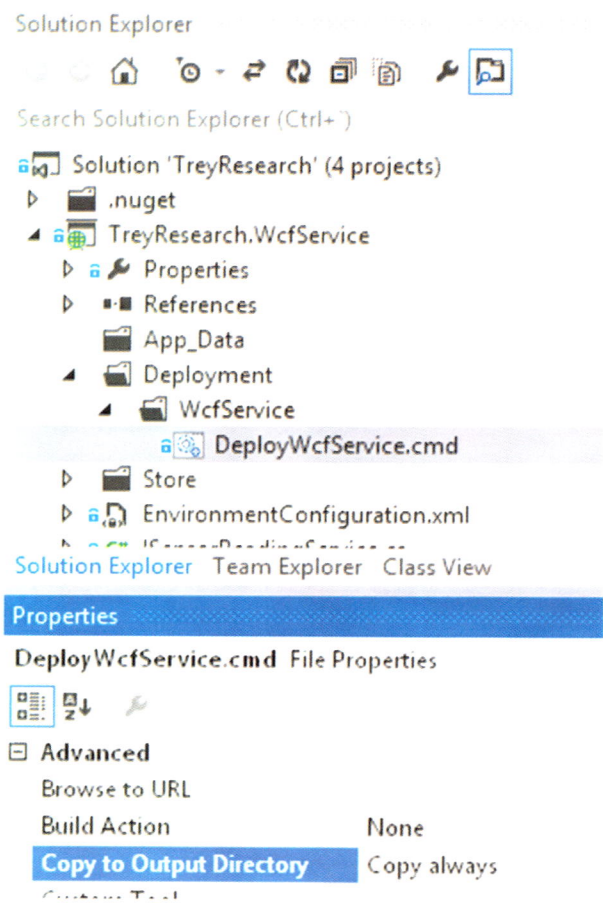

Up now, the pipeline always deploys the entire application, including the WCF web service, the WPF application, and the Windows Phone 8 app. This is a reasonable approach, at least for now, because the team is adding features that involve all three of them. The following screenshot shows that, during a deployment, all three deployment scripts are used.

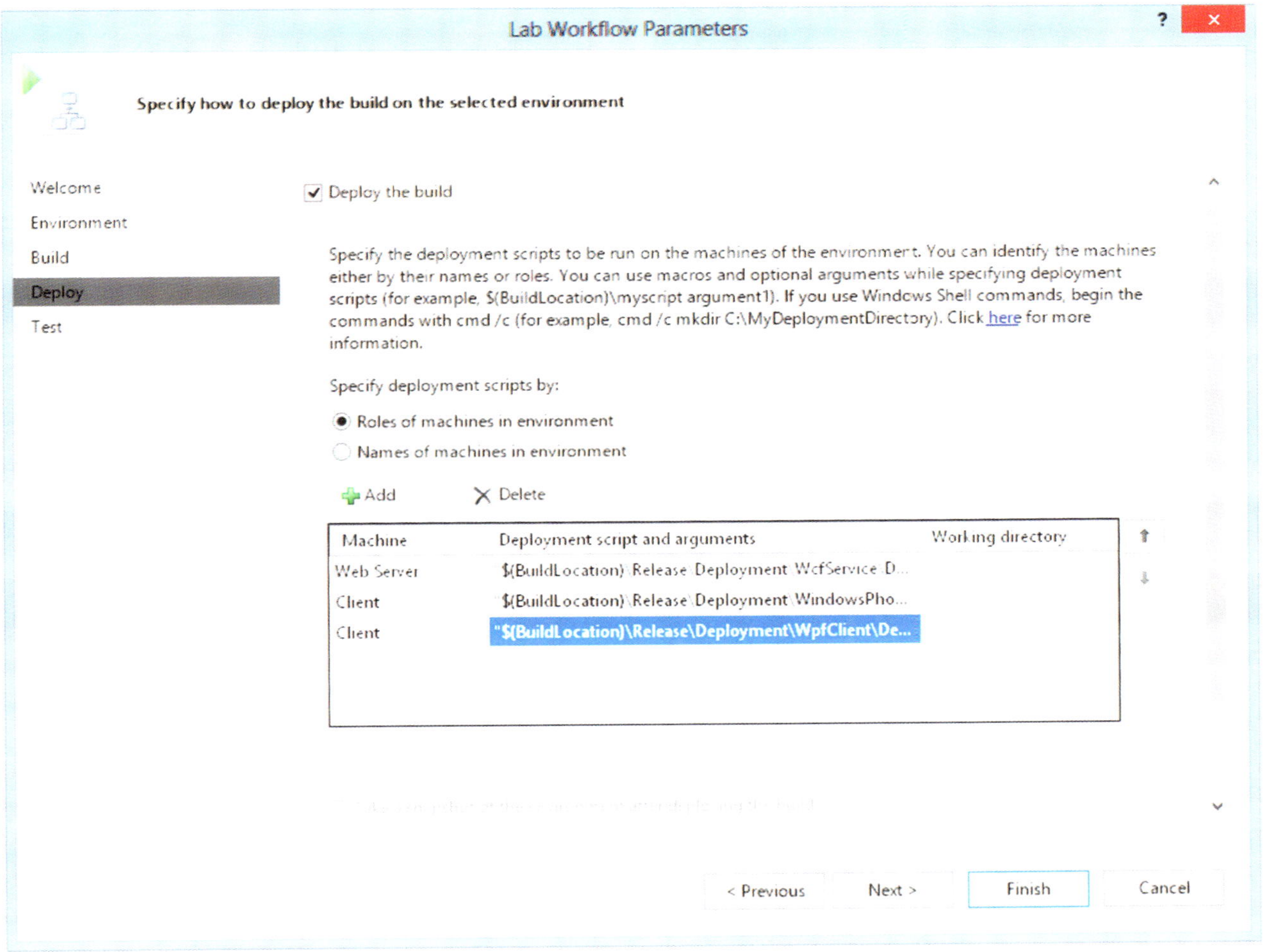

Lab Workflow Parameters

Specify how to deploy the build on the selected environment

Welcome
Environment
Build
Deploy
Test

☑ Deploy the build

Specify the deployment scripts to be run on the machines of the environment. You can identify the machines either by their names or roles. You can use macros and optional arguments while specifying deployment scripts (for example, $(BuildLocation)\myscript argument1). If you use Windows Shell commands, begin the commands with cmd /c (for example, cmd /c mkdir C:\MyDeploymentDirectory). Click here for more information.

Specify deployment scripts by:

● Roles of machines in environment
○ Names of machines in environment

➕ Add   ✕ Delete

Machine	Deployment script and arguments	Working directory
Web Server	$(BuildLocation)\Release\Deployment\WcfService\D...	
Client	$(BuildLocation)\Release\Deployment\WindowsPho...	
Client	"$(BuildLocation)\Release\Deployment\WpfClient\De...	

< Previous   Next >   Finish   Cancel

However, it would be straightforward to modify the pipeline so that it invokes only the script for the component that has changed. The team could divide each stage into three stages, one for each component. This effectively creates three different pipelines that are triggered only when there are changes to the stage's associated component.

Does Trey Research Have a DevOps Mindset

As we've seen, Trey Research has started to have planning meetings that involve people management, development, operations, and test. Also, there's now much more cooperation between team members. Because Trey Research is a small company, this is easier for them than it might be for people who work in large companies.

By the middle of the iteration, Jin is feeling hopeful.

Monday, August 26, 2013
We're starting to feel like a real DevOps team! Working together, we managed to automate the deployments and testing across all the pipeline stages. Now we can spend the rest of the iteration on building new features.

Here's the product backlog at the middle of the iteration.

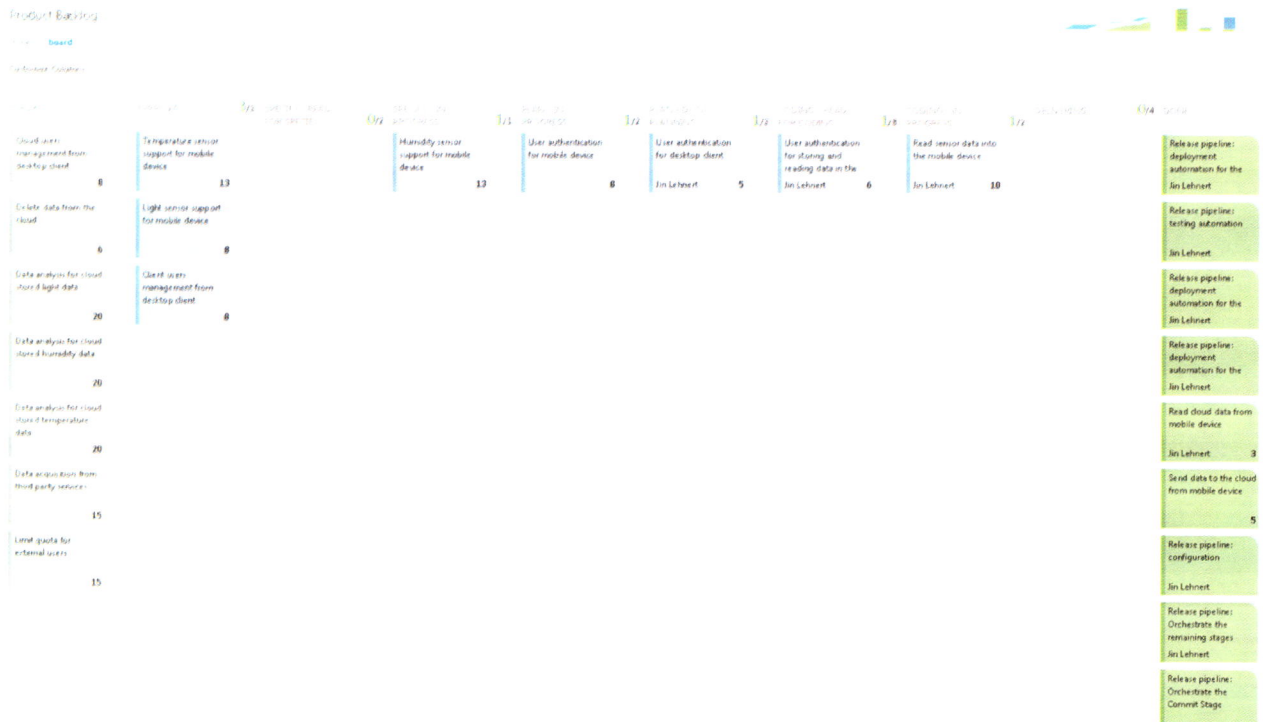

How Did Trey Research Create Automation Processes Early and Evolve Them

Trey Research began to automate their deployments and tests as soon as they finished orchestrating the pipeline. Fortunately, their application is a simple one, and they've been careful not to add lots of new features before the automation processes were in place.

What Tests Did Trey Research Choose To Run in Each Stage

For each stage, Trey Research runs tests that were appropriate but, for now, they run only a few types of tests.

- In the commit state, they only run unit tests.
- For the acceptance test stage, they run automated BVTs that validate the automated deployment. They also run automated acceptance tests that verify that the application still functions as it should after the deployment.
- For the release and UAT stages, the only run automated BVTs. Other types of tests are performed manually.
- In the future, they plan to add some new stages that are devoted to other types of testing, such as load and capacity tests.

Here are Jin's thoughts at the close of the iteration.

Friday, August 30, 2013

For the first time in the project, we managed to deliver all the forecasted work. It wasn't easy and we had a minor crisis near the end of the iteration. We realized that nobody was working on the bug that our user reported. The way the MSF Agile template handles bugs in TFS meant it didn't appear in the backlog or on the Kanban board, and those are our main tools for managing work. We had to put in a couple days of overtime and the work wasn't reflected in our backlog. We really need to find all these hidden work queues and make them visible.

Here's what the product backlog looks like at the end of the iteration.

Finally, Jin says this.

Friday, August 30, 2013

Now that the automation is in place, I feel like we can really say we use continuous delivery. We know that we can release the correct code whenever we want, and we're working faster, and with fewer problems. It all sounds great, but I expect our stakeholders will start making more demands. They know how quickly we can release new features, and they're going to want to start testing new ideas on users to see how they react. It's the right thing to do, but it means we can't just relax.

SUMMARY

In this chapter we talked about automating deployments and tests. Although it may be difficult to implement, automation has many advantages, such as freeing people from repetitive, monotonous, and error prone work. Automation also significantly improves the release process. By standardizing and automating your deployments, you remove many sources of problems, such as incorrectly configured environments. By moving from manual tests to automated ones, you increase the speed and reliability of your tests.

WHAT'S NEXT

In the next chapter, the team is celebrating because they now have a fully functional continuous delivery pipeline. They know that their release process has improved, but the problem is that they don't have any actual data that proves it. They need to learn how to monitor their pipeline so that they can collect all the data it generates and present it in a meaningful way. They also need to track some metrics that are particularly relevant to continuous delivery release process.

MORE INFORMATION

There are a number of resources listed in text throughout the book. These resources will provide additional background, bring you up to speed on various technologies, and so forth. For your convenience, there is a bibliography online that contains all the links so that these resources are just a click away. You can find the bibliography at: *http://msdn.microsoft.com/library/dn449954.aspx*.

For guidance about automatic deployments, see the ALM Rangers DevOps Tooling and Guidance website at *http://vsardevops.codeplex.com/*.

To learn about InRelease, which allows you to automate your deployments from TFS, see their website at *http://www.incyclesoftware.com/inrelease/*.

For guidance about creating builds, see the ALM Rangers Team Foundation Build Customization Guide at *http://vsarbuildguide.codeplex.com/*.

For guidance about using Microsoft Test Manager, see the ALM Rangers Test Release Management Guidance at *http://vsartestreleaseguide.codeplex.com/*.

For guidance about Visual Studio test features, such as CodedUI, see the ALM Rangers Visual Studio Test Tooling Guides at *http://vsartesttoolingguide.codeplex.com/*.

Another good testing reference is Testing for Continuous Delivery with Visual Studio 2012, which is available *http://msdn.microsoft.com/en-us/library/jj159345.aspx*.

For information about Active Directory Group Policy go to *http://support.microsoft.com/kb/816102*.

For information about System Center Service Manager and System Center Orchestrator, go to *http://www.microsoft.com/en-us/server-cloud/system-center/default.aspx*.

For information about Windows Installer, go to *http://msdn.microsoft.com/library/windows/desktop/cc185688(v=vs.85).aspx*.

For information about the WiX toolset, go to *http://wixtoolset.org/*.

For information about PsExec, go to *http://technet.microsoft.com/sysinternals/bb897553.aspx.*

For information about deployment orchestration models, see the "Deployment Scripting" section in Chapter of Jez Humble and David Farley's book, Continuous Delivery. To learn more about the preparation and management of useful test data suites, read Chapter 12. Learn more about the book at *http://continuousdelivery.com/*.

For information about SQL Server Data Tools, go to *http://msdn.microsoft.com/data/tools.aspx*.

For information about DBDeploy.NET, go to *http://www.build-doctor.com/2010/01/17/dbdeploy-net/.*

For information about Entity Framework Migrations, go to *http://msdn.microsoft.com/data/jj591621.aspx*.

The hands-on labs that accompany this guidance are available on the Microsoft Download Center at *http://go.microsoft.com/fwlink/p/?LinkID=317536*.

5 Getting Good Feedback

In the last chapter, the Trey Research team automated their release pipeline. They've come a long way. From a simple, largely manual pipeline that, in many ways, hampered their efforts rather than helped them, they now have a pipeline that ensures the quality of their software and has made releases, which were often chaotic and stressful, a far more predictable and repeatable process. Their pipeline has automated testing and automated deployments. It adheres to best practices for continuous delivery, such as using a single build that's deployed to many environments. The team knows that things are better. Their lives are easier and they're not staying at work until midnight, but they have no way of proving that the improvements they've made, which took time and money, are actually benefiting the business. Here's their situation.

This chapter is about how to really know how well your project is doing.

THE IMPORTANCE OF FEEDBACK

Many projects are managed without any concrete data that can help people make good decisions. A team might think that their work yields good results in a reasonable amount of time, but they don't have any actual information to prove it. Other teams spend valuable time tracking and analyzing either the wrong metrics, or metrics that give only a partial view of a situation.

The importance of good information that gives insight into a project can't be overemphasized. In general, we'll call this information feedback. Feedback is the most powerful tool a software development team can use to ensure that a project is progressing as it should.

The faster and more frequently feedback is available, the better a team can adapt to changes and anticipate problems. Feedback that is easy to generate, collect and act on can have a direct impact on a business. It helps you to focus your efforts in the right places if there are problems, and to create the new services and applications that your customers want.

The Importance of Good Communication

There are many ways to gather good feedback, and this chapter discusses some of them. However, the most important is to make sure that the people on the team talk to each other. Agile processes (for example, Scrum) prescribe periodic meetings such as standups where team members gather daily to learn what everyone's doing and where they can discuss issues. Agile processes also advocate retrospectives, where team members evaluate a project they've just completed to discuss what went well and what could be improved.

The Importance of Visibility

Agile processes stress good communication, not just in its verbal form, but also by using visual aids that encapsulate the status of a project. These aids are in a public place, where all team members can see them, and are easily understood. In Chapter 1 we discussed information radiators and used a traffic light as an example that many teams have adopted. The light quickly lets people know the status of the current build.

Another visual aid that is used in this guidance is the Kanban board. The Trey Research team uses these boards to understand the tasks that need to be done and their statuses. Examples of their Kanban boards are in several chapters of this guidance.

FEEDBACK, DevOps, AND CONTINUOUS DELIVERY

As you can imagine, both continuous delivery and DevOps rely on fast feedback to succeed. For DevOps, with its emphasis on collaboration, feedback is critical. DevOps stresses that everyone involved in a project must constantly communicate with each other.

However, because the focus of this guidance is the continuous delivery release pipeline, we're going to concentrate on how to get feedback from the pipeline itself. The three activities we'll discuss are:

- Generating feedback in the pipeline.
- Gathering feedback from the pipeline.
- Using metrics to evaluate the release process.

A fourth activity, acting on the feedback, depends very much on the situation you're in, and the type of feedback you're getting. This guidance shows how the Trey Research team reacts to various forms of feedback, much of it negative. Examples of acting on feedback include adding new features to the backlog, changing the way the team does code reviews, or even canceling a project. Although it's not possible to tell you how to act on the feedback you get, it is important to understand that acting on feedback as promptly and effectively as possible is the whole reason for generating and gathering it at all.

Generating Feedback

You can generate feedback from each stage of the pipeline. Generating feedback automatically is far better than trying to do it manually. After you've configured your pipeline, you collect data every time an instance of the pipeline runs, without having to do anything.

Gathering Feedback

At first glance, gathering feedback might seem trivial. It turns out, however, that it can often be difficult to retrieve, filter, manage ,and even uncover all the information that the pipeline generates. Two ways to gather information are to:

- Monitor the pipeline itself for information about running instances and the results from each stage.
- Monitor the application as it runs in the different environments.

Using Metrics

Another way to gather feedback is to use metrics. Metrics are so useful that they deserve a section of their own.

Continuous delivery pipelines use fast feedback loops to ensure that the code that is built works properly and is delivered promptly. It also uses feedback to validate that the code that you actually build is, among all the possibilities for what could be built, the best choice.

To validate such a choice, the most useful metrics for continuous delivery are those that assess the economic impact your choices have on the business. Many teams focus on metrics that actually aren't meaningful or that give only a partial view of a situation. For example, some teams measure the number of bugs they find in a project as a way to measure quality. However, if it takes three months to release a fix, simply knowing that the bug exists isn't that useful. The four metrics discussed in this chapter help organizations understand how effective their process is in its entirety, how often defects are discovered in the software, and how long it takes to remove those defects. These are the metrics.

- Cycle time, which is how long it takes between when you decide to make a change and when you deliver it. Cycle time is the most important metric for continuous delivery. This metric gives you a global view of your release process. It measures how the pipeline functions as a whole, and doesn't focus on the efforts of a particular discipline or organizational silo.
- Mean Time to Recovery (MTTR), which is the average time between when a problem is found in the production environment and when it is fixed.
- Mean Time Between Failures (MTBF), which is the average time between one failure in the production environment and the next failure.
- The defect rate, which is closely related to the MTBF and is the number of defects that are found per unit of time.

While these metrics aren't directly available in TFS, the information you need to calculate them is and is easy to retrieve. There are also other useful metrics that are directly available in TFS. Later in the chapter you'll find a brief discussion of these as well.

PATTERNS AND PRACTICES FOR GETTING GOOD FEEDBACK

This section discusses some patterns and practices that you can use to inform your approach to generating and gathering feedback from the pipeline, and for using metrics.

Automate the Generation and Gathering of Feedback

Just as with deployments and testing, automation makes it easier and more efficient to gather feedback from the pipeline. The pipeline can be configured to:

- Automatically generate information about running stages and the steps within them.
- Automatically gather that information and prepare it so that it is easily comprehensible.

Other than certain steps in the commit stage, such as the ones that build the binaries, and certain steps in the release stage, such as the ones that deploy to the production environment, everything else in the pipeline is there to provide some type of feedback. For example, you can get feedback about how a change to the code or to the configuration affects the application. The pipeline also tells you if the new code is ready to be delivered to your users.

If there is any information that you think will be useful, think about how you can configure the pipeline to generate that data. You may add a specific step to a stage, such as the code analysis step in the Trey Research pipeline. You can also add a new stage that generates a particular type of feedback. The acceptance test stage in the Trey Research pipeline is an example. It assesses whether the code behaves as expected after any change.

Of course, as always, if you can't automate at the moment, then run a manual step inside the pipeline. However, always make sure that you generate data that you can gather and present in an easily comprehensible way.

Design Software with the Operations Group in Mind

The way you architect, design, and code your software affects the quantity and quality of the feedback. Operations people need information about the applications they're responsible for maintaining and running. Software should be designed from the outset to provide information about an application's health, its status, and potential and immediate problems. From a DevOps perspective, involving operations people in the development process in order to include the correct instrumentation encourages collaboration between groups that frequently never communicate with each other.

The _Design for Operations website_ on CodePlex provides both a tool and guidance for creating highly manageable applications. One practice it advocates is to make business-related metrics available as well as information about an application's health and status. Business-related metrics might include the volume of data submitted to the application by web services at any moment, or the amount of money transferred as a result of financial transactions that occur within the application. Examples of information about the application itself are performance metrics, and the use of computing and network resources.

Another best practice is to use standard, well known instrumentation mechanisms that are generally familiar to operations people. They should be able to manage these mechanisms with standard monitoring tools such as Systems Center Operations Manager (SCOM). Examples of these instrumentation mechanisms include:

- Windows performance counters
- Windows event logs
- Windows management instrumentation
- Trace and log files

For more information about SCOM, see _System Center Operations – 2012 Operations Manager_.

Monitor the Pipeline

All the feedback that is generated and gathered is useless if you can't access it. Monitoring each pipeline instance allows you to can track the status of each change that is made and, in turn, the status of the application and the project as a whole.

Monitoring involves using a tool such as Build Explorer, which is available in Team Explorer. In terms of pipeline instances, the tool should provide information about:

- The instances of the pipeline that are running at any moment.
- The instances of the pipeline that have been completed.

For each of these instances, the tool should provide information about:

- The status of each stage, such as is it running, has it failed, has it partially succeeded, or has it succeeded entirely.
- The status of each step.
- Which manual stages and steps are ready to be run by a team member.

Monitor the Application

Monitoring the running application as changes are introduced across different environments is another way to obtain useful information. The feedback you get will be better if you have prepared your application according to the practices outlined in Design for Operations, however, even if you haven't, monitoring the application will alert you to potential and immediate issues. You can use this information to help you decide what needs to be improved or where to focus future development efforts.

Again, the amount and quality of the feedback is improved if you can automate how you monitor an application. Tools such as SCOM, once configured, can not only do the monitoring for you, but also generate alerts as a result of specific conditions. You can even provide this information to the team if you synchronize the alerts with Team Foundation Server. For more information, see *How to Synchronize Alerts with TFS in System Center 2012 SP1*.

Monitor the Work Queues

Long work queues, where tasks are inactive and wait for long periods of time before someone can address them, can cause many problems in a project. For example, cycle times can grow because tasks that must be performed before a feature is released to customers aren't being completed in a timely manner. A side effect is that it's also possible for people to begin to sacrifice quality in order to keep cycle times low.

Another issue is that long work queues can decrease the value of the feedback you're gathering because it isn't current. Timely feedback is only useful if tasks are closed quickly. Finally, tasks that sit in a queue for long periods of time can become outdated. The work they describe may no longer be applicable, or it may even have been invalidated.

Long queues can have negative effects on the team. For example, one commonly used metric is capacity utilization, which measures if people are working at full capacity. However, over emphasizing this metric can make queues longer. If people have no free time, then new tasks, which are probably at the end of the queue, don't merit immediate attention. Also, there's no incentive for people to complete tasks quickly if they're only judged by how busy they are. Measuring capacity utilization can discourage people from leaving some of their time unscheduled in order to react quickly to changes in the project. In addition, having to show that they're working all the time can put people under pressure, which is when they are most likely to make mistakes. A healthy project needs a balance between queue length and capacity utilization.

The problem is that it's difficult to make sensible tradeoffs if there's no information about the length of the work queues. Just like the pipeline, they need to be monitored. The first step is to decide what to track. Here are some important pieces of data.

Work in Progress

Work in progress is the first priority. If your team is working on an item, it belongs in a work queue.

Blocked Work

Blocked work also belongs in a queue. These are tasks that are waiting to be addressed by the team, have not yet been started, or are items for which the work has already begun but has been halted because there is a blocking issue. Blocked tasks can have long term harmful effects. They need to be explicitly included in the queue.

Hidden Work

Hidden work is very important. Hidden queues of work form when a team accepts tasks that are not explicitly tracked inside the queues used for regular project management. One very big problem with hidden work is that even though it can consume a great deal of a team's time, there's no way to make these efforts visible to management. A hidden task might be a special, urgent order from the CEO. It might be something that doesn't, initially, look like work but that keeps the team busy later. As soon as you detect hidden work, you should add it to the queue so that it's now visible and monitored.

How To Monitor Queues

Once you know what to track and your queues contain those tasks, you can track queue length and status. With TFS, you can use work items to create the queues and use work item queries to check the status and length of the queues. You can also create queues dedicated to a particular purpose, such as new features. If you use the MSF for Agile process template, then new features are **User Story** work items. Another example is a queue for defects. These are **Bug** work items in the MSF for Agile template. General work is described with **Task** work items.

In TFS, the Product Backlog work item query in the MSF for Agile process template is an example of a tool that helps you monitor and manage queues. The following screenshot shows an example of the results of such a query.

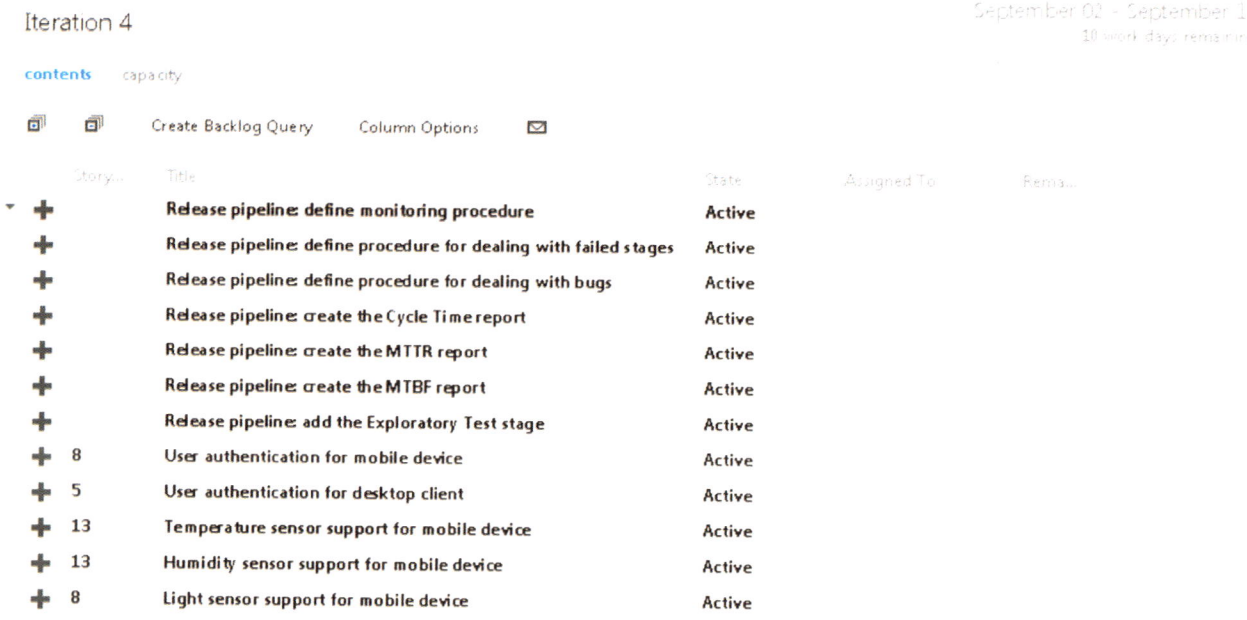

A convenient way to manage and monitor queues is the Cumulative Flow Diagram (CFD). The CFD provides a great deal of information in a single graph. It shows the number of items in the queue at each point in time, and differentiates them by their statuses. A CFD not only illustrates the status of the queues, but also gives insight into the cause of some problems, and can even help you to discover potential issues. Another good point about a CFD is that the information is up to date because the queues are shown in their current state.

The following screenshot shows the CFD for the Trey Research project.

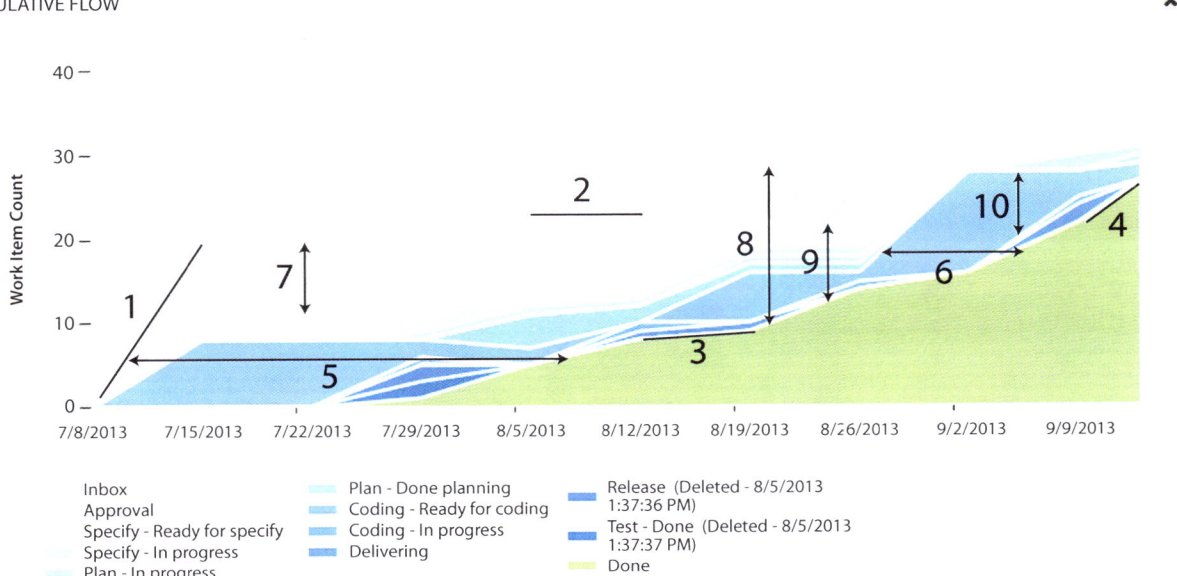

Here are some of the things you can learn by analyzing a CFD. Numbers in the following text correspond to numbers in the graph.

- **The rate at which items enter and leave queues, as well as variations to that rate**. The slope of the curves provides this information. For example, if you compare the slope of lines (1) and (2), you can conclude that during the first week of the project, as many as 20 new items were added to the backlog. Later, around the fifth week of the project, no new items were being added. Again, by examining the slopes, you can see that, by the last week, the team was delivering items at a higher pace (4) than four weeks earlier (3).

- **The time items spend in each queue (or sets of queues)**. The width of the stripes provides this information. One of most important times to track is the lead time (5), which represents the total time spent on all activities until the item is done. You can see that, at the beginning of the project, the lead time was about 5 weeks. This means that it would take a new item approximately five weeks to be released. The cycle time (6), which tracks the time spent by the development team on new items, was around one week at the point marked in the graph.

- **The number of items in each queue (or in sets of queues)**. The height of the stripes provides this information. For example, you can see the backlog size at the beginning of the third week (7), and how it decreases later. You can also see that the amount of unfinished work (8) has been steady for the duration of the project. This is also true of the amount of work in progress (9). These are probably consequences of there being work-in-progress limits that are enforced by the Kanban board so that the team doesn't accept more tasks than it can handle. You can also track specific queues. You can see that the number of items waiting in the **Ready for coding** queue (10) began to decrease around the ninth week, after being unusually high during the preceding weeks. It's possible that the team, by constantly analyzing the CFD and the Kanban board took some action that addressed the problem.

All this information is easily generated simply by keeping the work items in TFS up to date.

Use Metrics

William Thomson, the 1st Baron Kelvin, is best known for his work on the laws of thermodynamics, and for determining the value of absolute zero. His statement on the importance of measurement is also applicable to software development.

I often say that when you can measure what you are speaking about, and express it in numbers, you know something about it; but when you cannot express it in numbers, your knowledge is of a meagre and unsatisfactory kind; it may be the beginning of knowledge, but you have scarcely, in your thoughts, advanced to the stage of science, whatever the matter may be.

This section discusses four important metrics for continuous delivery: cycle time, MTTR, MTBF and the defect rate.

What Is Cycle Time

Cycle time is the development (or implementation time). This is in contrast to lead time, which includes all the activities that occur until an item is completed. For example, lead time includes the time spent by stakeholders deciding if a feature should be implemented.

Why Track and Analyze Cycle Time

There are many reasons why tracking and analyzing cycle time is a valuable activity. Here are some of them.

- Cycle time measures the effectiveness of the entire pipeline.
- Cycle time helps you to identify wasteful activities and eliminate them, or to improve the way activities are currently performed.
- Cycle time allows you to uncover bottlenecks in the release process.
- Cycle time gives insight into how changes to the release process and even specific decisions affect the time it takes to deliver the software to users.
- Cycle time helps you to measure the predictability of the release process, and, after the process stabilizes, can help you to make better forecasts about when you can deliver your software to users.

How Do You Measure Cycle Time

In order to measure cycle time, you need to record when the implementation starts, and when the change is available to your users. The amount of time between these two moments is the cycle time for that change.

Again, cycle time is a way to measure and improve the implementation process. What implementation means can differ from one organization to another but, in simple terms, it would be fair to say that it is the set of activities performed by the development team.

The units to use for measuring cycle time depend on how long it takes, on average, for a team to deliver its new software. Many teams find that the day is a good unit to use.

You can plot the cycle time of each change on a graph so that you can see the trend and also identify and investigate edge cases. When a project has been active for a long time, or if there's a large number of changes that are released, it can be useful to group these changes. For example, you might want to obtain the average cycle time over some range of weeks, in order to see how the metric changes.

How Can You Use Cycle Time to Improve the Release Process

As a rule of thumb, the shorter the cycle times the better, so the trend line should be descending. This means that you are eliminating wasted efforts and bottlenecks and your release times are therefore growing shorter. Make sure that minimizing the cycle time doesn't become a goal in and of itself. The goal should always be to deliver valuable software that is good for your customers. If reducing cycle time becomes the objective, it can come at the cost of sacrificing the quality of the software or by taking shortcuts. The way to lower cycle times is to optimize the release process by automating repetitive tasks and by getting rid of unnecessary ones.

After spending some time improving the release process, cycle times become more predictable and the values will probably fluctuate around a fixed range. This might mean that you should look for improvements in new areas. Of course, you should try to maintain this stable pattern and not have the values increase again.

There are situations where increasing cycle times are acceptable. For example, if you're focusing on improving the release process itself, perhaps by implementing new stages for the pipeline or by adding automation, your cycle times will probably increase.

A disadvantage of cycle time is that it's a lagging metric, in contrast, for example, to the CFD, which shows data in its current state. You can't measure the cycle time for an item until the work is complete. It's a good idea to track cycle time in conjunction with the CFD.

What is MTTR

MTTR is the average time between the moment a problem is found in the production environment and the moment that the problem is fixed. Production bugs are the focus because bugs found and fixed during the development process don't have a direct impact on the business.

MTTR is also known as Mean Time to Resolve and Mean Time To Repair. Within the Information Technology Infrastructure Library (ITIL) it's named Mean Time to Restore Service (MTRS). (ITIL is a set of practices widely used for IT service management.)

Why Track and Analyze MTTR

MTTR is similar to cycle time. Tracking this metric yields the same types of benefits, such as the ability to identify and remove bottlenecks. The difference is that MTTR is related to the ability to resolve defects and deliver fixes rather than implement and deliver new features. You can think of MTTR as a special case of cycle time.

Measuring MTTR independently of cycle time is a good idea because most teams are particularly interested in bugs. For example, some teams have a zero defect policy, where any bug is either resolved immediately or discarded. In this situation, it's very useful to know the average time needed to fix a problem. A low MTTR also points to a better experience for end users and stakeholders. Generally, a low MTTR means that customers encounter quick resolutions to problems. For stakeholders, the sooner a bug is fixed, the less impact it has on the business.

How Do You Measure MTTR

To measure MTTR, you do the same as you would for cycle time, but use defects as the basis of the measurement instead of new features. For each defect, you record when the problem was found and when it was fixed, in the production environment. The amount of time between these two moments is the time to recover for that defect. The MTTR represents the average time that it takes to recover from a defect.

The units for measuring MTTR depend on how long it takes, on average, for a team to fix production bugs. This can depend on the policy of the organization about defects, and the number of critical bugs as opposed to those of low importance. For some teams, it might take days, but for other teams a better unit is the hour.

As with cycle time, you can plot MTTR on a graph, so that you can see the trend and also identify and investigate edge cases. When a project has been active for a long time, or if there is a large number of fixed bugs, it can be useful to group them. For example, you might want to obtain the average MTTR over some range of weeks, in order to see how the metric changes.

To improve the development process, calculate the MTTR using only the time actually spent fixing bugs. Don't include other activities such as triaging bugs. To improve the overall process of fixing bugs, calculate the MTTR and do include other, related activities such as triaging bugs.

How Can You Use MTTR to Improve the Release Process

Once again, the answer is similar to cycle time. A low MTTR is better, so the trend should be descending. A larger MTTR is acceptable in some situations, such as when you're improving the release process but (hopefully) not negatively affecting the business.

What Is MTBF and the Defect Rate

MTBF is the average time between the moment one problem is found in the production environment and the moment that the next problem is found in the production environment. As with MTTR, MTBF only includes production bugs because they are the bugs that can have a direct impact on the business. MTBF is closely related to the defect rate. They are inverses of each other, so MTBF = 1 / Defect rate. In other words, the defect rate is the number of defects found for each unit of time.

Why Track and Analyze MTBF and the Defect Rate

These metrics help you to keep track of the quality of your software. If MTBF decreases, (or the defect rate increases), it can signal a quality control policy that is too lax or is being ignored. Poor quality control can have a direct impact on the business. When customers keep finding bugs they become frustrated and lose confidence. It also means that the application isn't functioning properly, which can have a negative economic effect on the business.

The MTBF, the defect rate, and the cycle time are closely related. Typically, more defects means there is less time to spend on new features, so you may see an increase in cycle times if the MTBF decreases.

There is a close relationship between MTBF and MTTR as well. Together, these two metrics indicate the overall availability of the application.

How Do You Measure MTBF and the Defect Rate

To measure MTBF, you record when each defect is found and calculate the average time between defects. The units of measurement depend on how often bugs are found, but for most teams, either the day or the hour should be suitable.

For the defect rate, you count the number of defects in the system. The unit of measurement is the number of defects per unit of time (for example, the number of defects per day or per week).

Plotting the MTBF and the defect rate lends a better understanding of these metrics. You can see the trends and examine edge cases. Even though one metric can be derived from the other, it's still valuable to track them independently. MTBF and the defect rate provide different views of the same information, so they complement each other.

When a project has been active for a long time, or if there is a large number of bugs, it can be useful to group the bugs. For example, you might want to obtain the average MTBF over some range of weeks, in order to see how the metric changes.

How Can You Use MTBF and the Defect Rate to Improve the Release Process

MTBF should be as large as possible. The more time that passes between one defect and the next one, the better. The defect rate should be as small as possible. A smaller MTBF (or a larger defect rate) indicates that something is wrong with how quality is assessed in the development process.

Other Useful Metrics

There are many other metrics that are useful for providing feedback about a project. Every time you create a new team project, the **New Team Project** wizard generates a set of standard reports, such as velocity and burndown rates, depending on the process template that you select. These reports are available from Team Explorer. Every team project has a **Reports** node, where you will find the reports that have been generated. For more information about standard TFS reports, see *Create, Customize, and Manage Reports for Visual Studio ALM*. You can use tools such as Build Explorer to get information about automated builds. If you are interested in data about your code, you can use the **Analyze** menu that is included with Visual Studio Premium or Visual Studio Ultimate. In terms of continuous delivery, however, while these are certainly useful metrics, they aren't mandatory.

THE TREY RESEARCH IMPLEMENTATION

Now let's take a look at how Trey Research is implementing these patterns and practices. When we left the team, they'd finished automating their pipeline, and now had a fully functional continuous delivery release pipeline. They still have some unsolved problems, though. Here are the ones they're going to address in this iteration.

Issue	Cause	Solution
For each change to the code, they don't have an easy way to know if the change meets all the conditions that make it eligible for release to production.	They don't monitor the pipeline in a way that makes it easy to know what happens to a change in each pipeline stage.	Use Build Explorer and the corresponding Build section inside the TFS team project Web Access site.
They don't have enough information to help them make good decisions during the development process.	They are missing some key metrics.	Start to track cycle time, MTBF, and MTTR.

The first solution involves using tools that they already have as part of TFS. The second solution involves learning how to create custom reports. The following pipeline diagram includes these activities.

Pipeline with Monitoring and Metrics

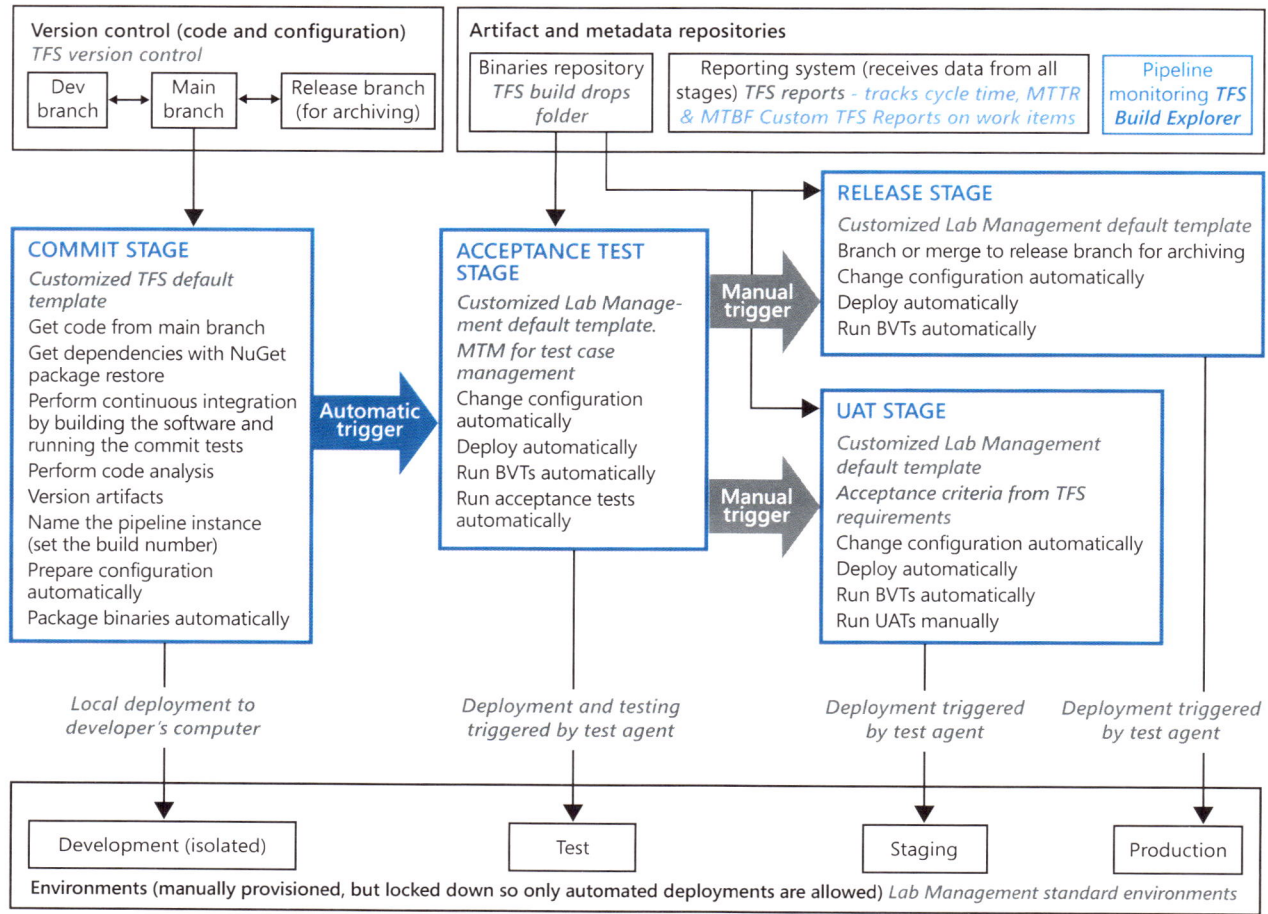

The rest of this chapter explains what the Trey Research team did. As a general description, we can say that they focused on improving transparency and visibility. They've learned how to monitor the pipeline and to use automation in order to gather feedback. They've learned how to use metrics, and how to present the feedback they've gathered. In particular, they've begun to track cycle time, MTTR, and MTBF so that they can evaluate how the changes they're making impact the quality of their software and the business. First, let's get back to Jin's story.

Jin's Story

Here are Jin's feelings at the outset of the iteration.

> **Monday, September 2, 2013**
> Everyone on the team is euphoric after we got our pipeline working. Sure, we deserve a party, but we're not close to being finished. My problem is that we can't measure what we've done, and if we can't do that how do we prove to management it was all worth it? So, we're going to make sure we get feedback from the pipeline and we're going to start using metrics.
> Another good thing is that the team is really feeling involved. Iselda's happy so many of the tests are automated, and she's even suggesting we add an exploratory testing stage. Raymond is happy because releases aren't keeping him up until 03:00. Paulus is happy because he's getting quick feedback on his new features, and isn't chasing bugs that don't exist. Even Zachary is happy. The app works, people like it, and we're making money. Hey, we're all happy.

Here's the Trey Research backlog for iteration 4.

Iteration 4

September 9 — September 13
10 work days remaining

contents   capacity

Create Backlog Query   Column Options

	Story	Title	State	Assigned To	Remai...
+		Release pipeline: define monitoring procedure	Active		
+		Release pipeline: define procedure for dealing with failed stages	Active		
+		Release pipeline: define procedure for dealing with bugs	Active		
+		Release pipeline: create the Cycle Time report	Active		
+		Release pipeline: create the MTTR report	Active		
+		Release pipeline: create the MTBF report	Active		
+		Release pipeline: add the Exploratory Test stage	Active		
+	8	User authentication for mobile device	Active		
+	5	User authentication for desktop client	Active		
+	13	Temperature sensor support for mobile device	Active		
+	13	Humidity sensor support for mobile device	Active		
+	8	Light sensor support for mobile device	Active		

Jin's next entry tells us how things turned out at the end.

> **Friday, September 13, 2013**
> We implemented the new metrics. We're tracking cycle time, MTBF, and MTTR. Cycle time's still unpredictable because of all the time we spent on the pipeline but the good news is that all the trend lines are heading in the right direction and that's something we can show to management.

Here's the Trey Research cycle time, over a series of weeks.

Average Cycle Time per week, in days

Week	Average Cycle Time in days
30/2013	10
32/2013	11
33/2013	10
34/2013	5
35/2013	11
36/2013	3
37/2013	9

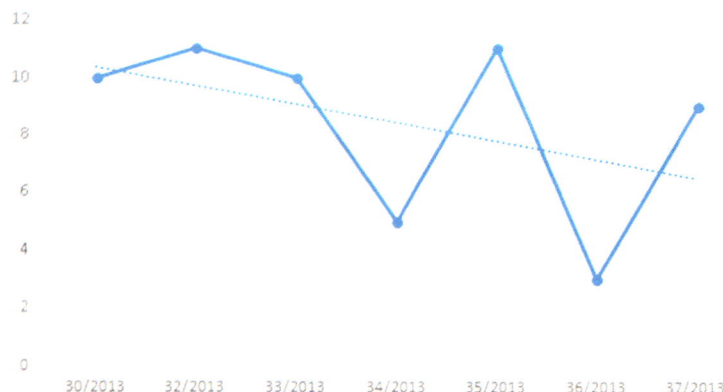

Here's their CFD.

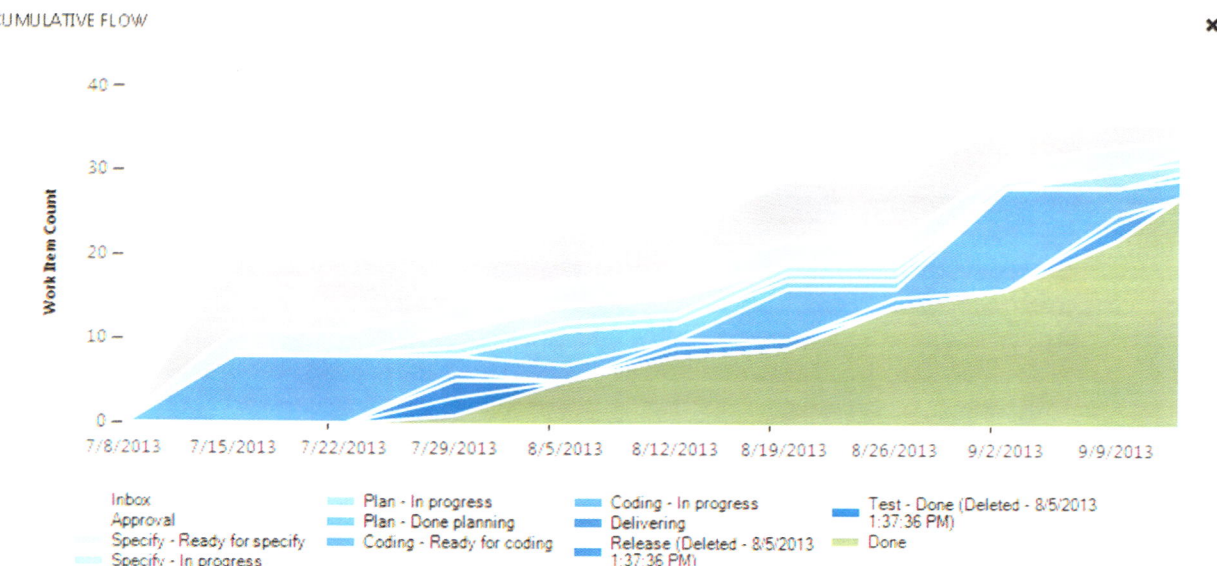

For more information about how to generate and interpret the Trey Research reports, see *Lab 4.2: Metrics for Continuous Delivery in TFS*.

HOW DID TREY RESEARCH ADD MONITORING AND METRICS

This section discusses how Trey Research implemented monitoring and metrics, and the best patterns and practices that they followed. For a step-by-step description, see the group of labs included in *Lab04-Monitoring*.

How Is Trey Research Automating the Generation and Gathering of Feedback

When they designed the pipeline, the Trey Research team made sure that stages and/or steps were in place for the types of feedback they needed to ensure the quality of their software. They use different mechanisms to generate and gather feedback. Some of it is done by using the logging and tracing mechanisms that the pipeline provides. The following screenshot shows the **WriteBuildMessage** workflow activity the team uses to generate some feedback about the stages that the pipeline is about to trigger.

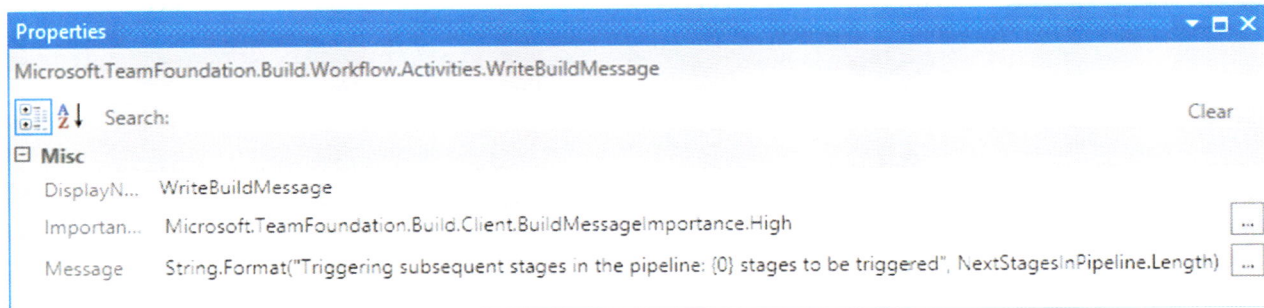

The automated tests that the team uses also provide feedback in the form of test results that can be read and analyzed after each testing stage has finished. There are also some steps in some stages that generate specific types of feedback. An example is the code analysis that is always performed in the commit stage.

How Is Trey Research Designing for Operations

The team has been so busy implementing the orchestration, automation, and monitoring of their release pipeline that they haven't done this. It's in their backlog and when they have time, they'll address this issue.

How is Trey Research Monitoring the Pipeline

The Trey Research team uses Build Explorer and the corresponding area inside the TFS team project Web Access site.

As well as using Build Explorer, the Trey Research team use the alerts system in TFS to receive prompt notification about important events that occur within the pipeline. The following screenshot shows an example of how they've set up an alert in TFS that uses email to warn them if any of the stages in a pipeline instance fail.

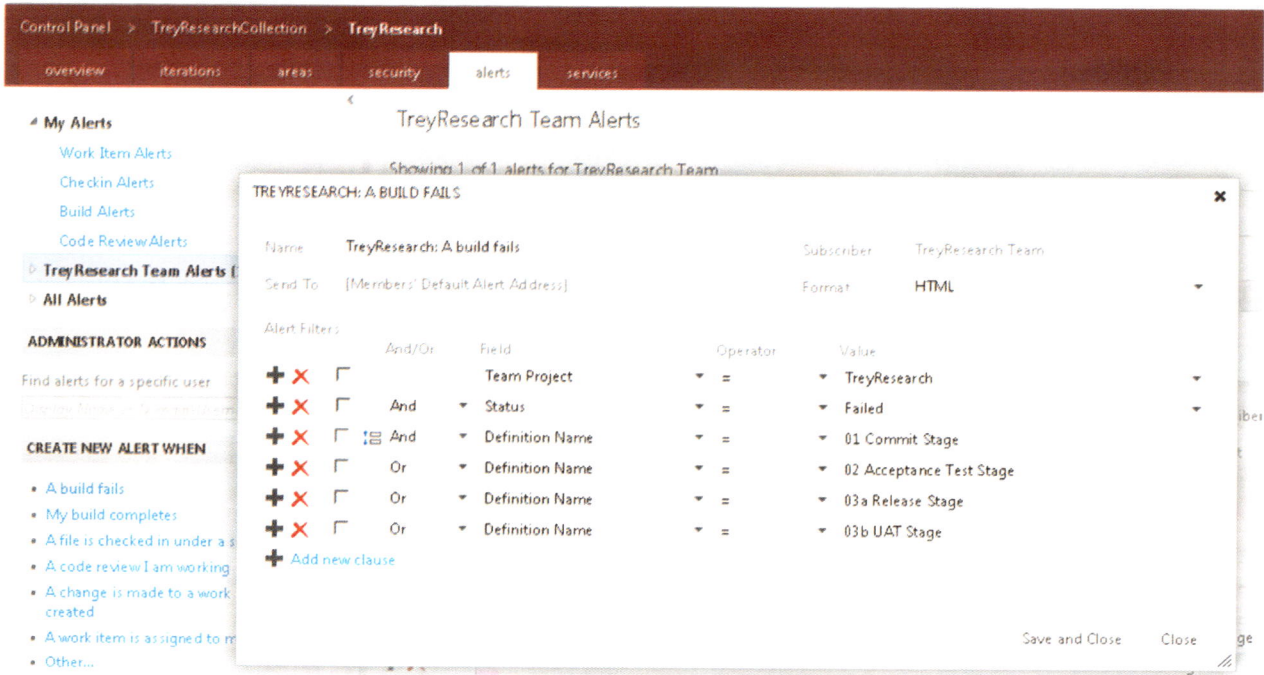

How Is Trey Research Monitoring the Application

Just as with the designing for operations, the team hasn't had the chance to set up application monitoring. Again, this is on their backlog.

How Is Trey Research Monitoring the Work Queues

The team uses TFS to enter all their tasks as work items, to track the work items, and to get information about them by using queries. For more information, see *Process Guides and Process Templates for Team Foundation Server*.

They also use the CFD that is available through the TFS team project Web Access site, as well as the TFS Kanban board. For more information about Kanban boards, see *Manage your backlog with the Kanban board*.

How Is Trey Research Tracking and Analyzing Cycle Time

The team uses TFS work items to list and manage features to be implemented. They use the activated and closed dates of the **User Story** work items to calculate the cycle time for each work item. They prepare a custom TFS report that shows the trend over time. (Although cycle time is shown in the CFD, the custom report is easy to prepare and it provides a more detailed view.)

How Is Trey Research Tracking and Analyzing MTTR

The team uses TFS work items to list and manage production bugs. They use the activated and closed dates of the **Bug** work items to calculate the MTTR. They prepare a custom TFS report that shows the trend over time. For complete details on how to track the MTTR, generate a custom report, and interpret the results, see *Lab 4.2: Metrics for Continuous Delivery in TFS*. In the future, the Trey Research team plans to improve how they track this metric by distinguishing between production bugs and other types of bugs and to filter using the appropriate classification.

How Is Trey Research Tracking and Analyzing MTBF and the Defect Rate

The team uses TFS work items to list and manage production bugs. They use the activated and closed dates of the **Bug** work items to calculate the MTBF, and they only use bugs that are already closed to make sure they don't include bugs that are invalid. They prepare a custom TFS report that shows the trend over time. They haven't starting explicitly tracking the defect rate, but because this is a standard TFS report, they plan on using it soon.

For complete details on how to track the MTBF, generate a custom report, and interpret the results, see Lab 4.2: Metrics for Continuous Delivery in TFS. In the future, the Trey Research team plans to improve how they track this metric by distinguishing between production bugs and other types of bugs and to filter using the appropriate classification.

The team also uses the standard Bug Trends Report to help track the rate at which they discover and resolve bugs. For more information, see *Bug Trends Report*.

THE REVISED VALUE STREAM MAP

After the team spent some time gathering data, they completed the value stream map they created in Chapter 3 by replacing the placeholders with times. They used the CFD and cycle time reports to get the approximate values. Here is what the new value stream map looks like.

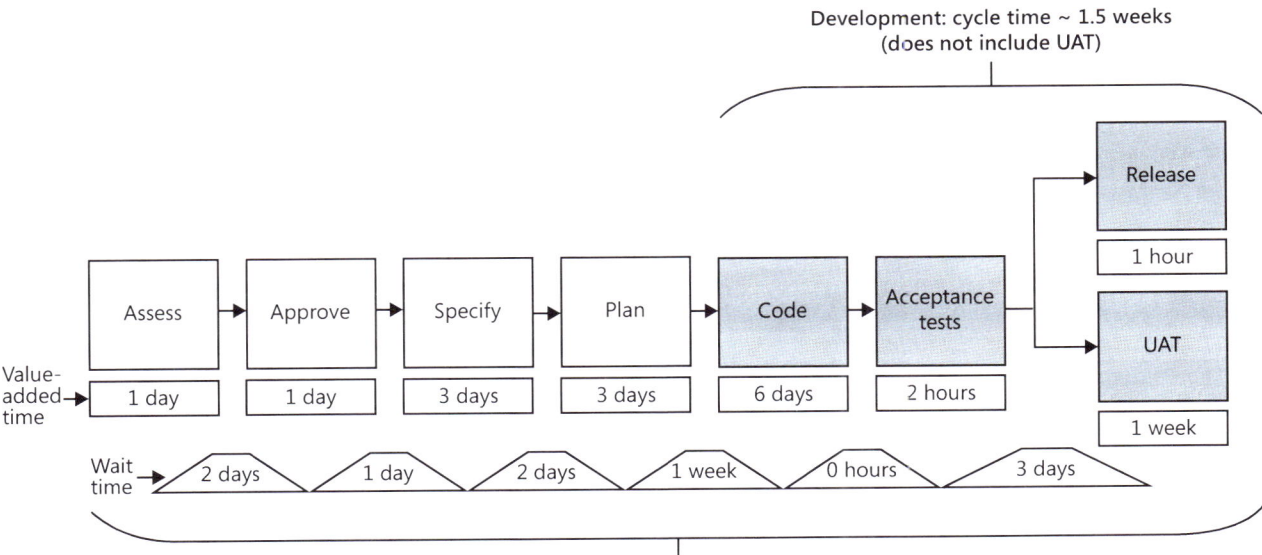

The team didn't include the UAT time in their calculations because it depends on when end users can dedicate some time to testing the application. Also, they now have high confidence in their acceptance tests, so they release the application in parallel to when it enters the UAT stage.

After all their efforts, they found that the cycle time is almost half what it was when they began. Their lead time has decreased by 1.5 weeks. Orchestrating and automating the pipeline, as well as following the best practices for continuous delivery, has helped them to dramatically reduce both their value-added time and their wait time. The most extreme example is the wait time between the code and acceptance test activities. In fact, there is no longer a wait time because the transition between the two activities is done automatically, inside the pipeline, when the commit stage automatically triggers the acceptance test stage.

Summary

This chapter discussed ways to get good feedback about your project. There are some patterns and practices you can follow, no matter what technology you use. Some of them include using automation, monitoring the pipeline, the application, and work queues. You also learned to interpret a CFD, which encapsulates a great deal of information about a project. The chapter also stresses the importance of metrics. In particular, the success of a continuous delivery pipeline is best shown by tracking cycle time, MTTR and MTBF.

More Information

There are a number of resources listed in text throughout the book. These resources will provide additional background, bring you up to speed on various technologies, and so forth. For your convenience, there is a bibliography online that contains all the links so that these resources are just a click away. You can find the bibliography at: *http://msdn.microsoft.com/library/dn449954.aspx*.

The book *Principles of Product Development Flow* by Donald G. Reinertsen has a great deal of information about how to monitor and manage queues. Although it covers all types of product development, the principles it discusses also apply to software development. For more information, see the Reinertsen & Associates website at *http://www.reinertsenassociates.com/*.

There is another approach to calculating MTBF that uses a slightly different method. It measures the time between the moment when a defect is fixed to the moment when a new defect appears. By defining the metric this way, you learn the average time the application is available. For more information, see the Wikipedia article about mean time between failures at *http://en.wikipedia.org/wiki/Mean_time_between_failures*.

The Design for Operations website at *http://dfo.codeplex.com/* provides both a tool and guidance for creating highly manageable applications.

For information about standard TFS reports, see Create, Customize, and Manage Reports for Visual Studio ALM at *http://msdn.microsoft.com/library/bb649552.aspx*.

For information about SCOM, see System Center Operations – 2012 Operations Manager at *http://technet.microsoft.com/systemcenter/hh285243*.

For information about how to use alerts with TFS, see How to Synchronize Alerts with TFS in System Center 2012 SP1 at *http://technet.microsoft.com/library/jj614615.aspx*.

For information about managing work items, see Process Guides and Process Templates for Team Foundation Server at *http://msdn.microsoft.com/library/hh533801.aspx*.

For information about using TFS Kanban boards to manage your backlog, see Manage your backlog with the Kanban board at *http://msdn.microsoft.com/library/vstudio/jj838789.aspx*.

For information, about using standard bug trend reports, see Bug Trends Report at *http://msdn.microsoft.com/library/dd380674.aspx*.

The hands-on labs that accompany this guidance are available on the Microsoft Download Center at *http://go.microsoft.com/fwlink/p/?LinkID=317536*.

6   Improving the Pipeline

This chapter offers suggestions for ways to further improve the continuous delivery pipeline and its associated components and practices.

The Trey Research Team has come a long way. They have a continuous delivery pipeline that they can monitor for valuable feedback. They're also tracking metrics that let them know how well they're doing, in terms of adding value to the business. Are they finished? Of course not. There are still problems and there are many improvements that they can make.

This chapter offers suggestions for ways to further improve the pipeline and its associated components and practices. For example, there's a discussion of possible branching strategies. While none of these areas are discussed in great depth, we hope there's enough information to help you identify areas that can benefit from some changes, and some practical guidance on how you might implement those changes.

PATTERNS AND PRACTICES FOR IMPROVING THE PIPELINE

Here are some patterns and best practices to follow to further improve a continuous delivery pipeline, no matter which specific technologies you use.

Use Visuals to Display Data

Your first reaction to the feedback you see may be that it's incomprehensible. For example, if you use Build Explorer to monitor your pipeline there's a great deal of useful information, but you may want to make it easier to understand. Information becomes much more accessible if you implement a pipeline solution that automatically creates some visuals that present data in an organized and easily understandable format.

An especially helpful visual technique is a matrix view that shows the status of multiple pipeline instances. Another useful visual is a flowchart view, which shows the status of one pipeline instance, the sequence of stages, and the dependencies between them. An example of a dependency is that the acceptance test stage requires the commit stage to succeed before it can be triggered. Finally, another improvement is to have some easily identifiable way to open the build summary page for a stage, in order to make it easy to get more detailed information.

The following illustration is an example of a matrix view. With it, you can easily assess the status of multiple pipeline instances.

Pipeline instance	Commit stage	Acceptance test stage	Release stage	UAT stage
0.0.133.12	Completed	Failed		
0.0.133.13	Completed	Completed	Completed	Completed
0.0.133.14	Completed	Completed	Pending Trigger now	Pending Trigger now
0.0.133.15	Failed			
0.0.133.16	Running			

The following illustration is an example of a flowchart view. You can see the status of each stage of a particular pipeline instance.

Pipeline instance: 0.0.133.11

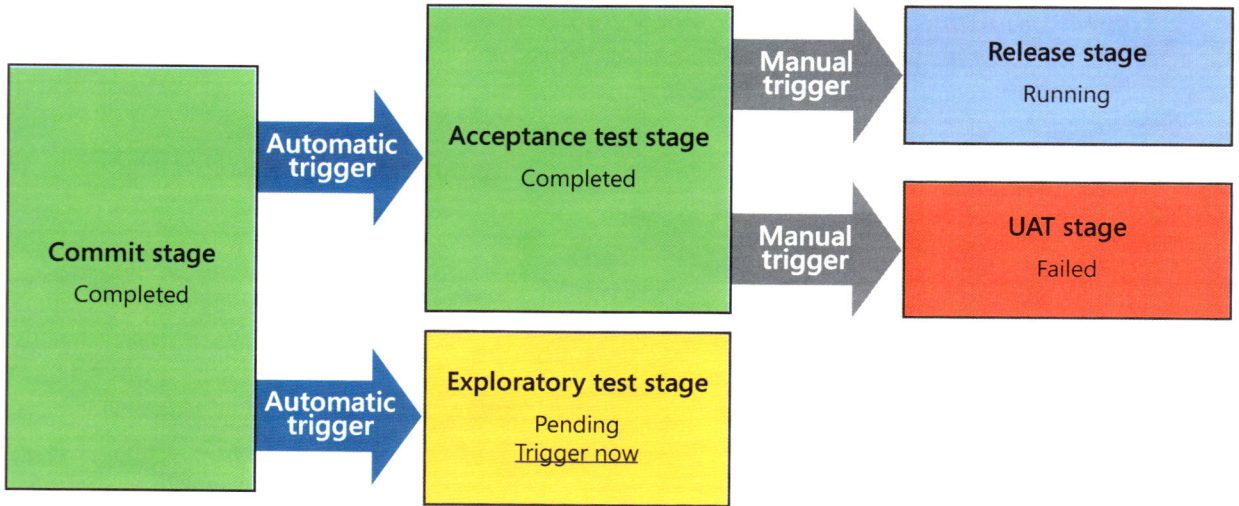

Choose a Suitable Branching Strategy

Coordinating the changes that multiple developers contribute to the same code base can be difficult, even if there are only a few programmers. Code branching is the most widely adopted strategy, especially since distributed version control systems such as Git have become popular. (For information on how to use Git with TFS, see *Brian Harry's blog*.)

However, branching is not always the best strategy if you want to use continuous delivery. In the context of a continuous delivery pipeline, it's unclear how to work with merge operations. Best practices such as build only once, or propagate changes automatically are harder to adopt if merges across several branches are required. You can't build only once if you merge several branches. Each time you merge, the code has to be rebuilt before performing validations on that branch.

You also can't propagate changes automatically if, during the propagation, you have to deal with a merge operation, which many times isn't completed automatically because of conflicts. You can read more about branching in the context of continuous delivery in Jez Humble's article *DVCS, continuous integration, and feature branches*.

In terms of continuous delivery, a better approach is to get rid of branching and use feature toggles instead. A feature toggle makes it possible to either show users new features or not. In other words, you can toggle a feature's visibility.

With feature toggles, everyone works from the main source trunk and features can be deployed to production even if they're not finished, as long as they're hidden. Feature toggles are often implemented with configuration switches that can easily be changed to activate or deactivate specific features at run time. By using feature toggles, your code is always in a production ready state, and you only have to maintain a single line of development. Feature toggles also enable other continuous delivery practices such as A/B testing and canary releases, which are covered later in this chapter. A good place to start learning about feature toggles is *Martin Fowler's blog*.

Many teams feel that feature toggles are difficult to implement and so prefer branching. If your team feels the same way, there are some best practices to follow if you practice continuous delivery.

Use Feature Branching

Use feature branching, by which we mean that you should set up a different development branch for each new feature. This is in preference to any other branching strategy, such as having a unique development branch. These strategies make it even more difficult to follow best practices for continuous delivery, such as always having a version of your code that is ready to go to production.

Feature Branches Should Be Short Lived

Feature branches should be as short lived as possible. Don't let changes stay too long in a branch. The longer changes are unmerged, the more difficult the merge operation will be. By short lived, we mean that a branch typically exists for a few hours, and never longer than a few days.

Keep Features Small

To have short-lived feature branches, make sure that the feature is small enough so that it can be implemented in a few days, at most.

Define and Enforce a Merge Policy

Define a merge policy and enforce it. Small changes should be frequently integrated with the main branch, and from there to any other development branches. Any time a feature is implemented, the corresponding development branch should be merged into the main branch, and the development branch should be discarded. Any time the main branch incorporates changes (because of a merge operation), these changes must be propagated to other active development branches, so that the integration is done there, to minimize integration problems with the main branch.

Don't Cherry Pick

Cherry picking means that you merge only specific check-ins, instead of all the changes, made to a branch. Always merge from the root of the source branch to the root of the target and merge the entire set of changes. Cherry picking makes it difficult to track which changes have been merged and which haven't, so integration becomes more complex. If you find that you must cherry pick, then consider it a symptom of a feature that is too broadly defined, and whose corresponding branch contains more items than it should.

Make the Main Branch the Pipeline Source

Always make the main branch the source for the pipeline. The pipeline must build once and changes should be propagated automatically as much as possible. The easiest way to accomplish this is to avoid merge operations in the context of the pipeline. Make the main branch the source for changes that are sent through the pipeline. The main branch contains the code that is complete and ready for production after it passes all the validations performed by the pipeline stages.

Another option is to have different pipelines that support different branches, where the pipeline would validate the code in each development branch just before the code is merged with the main branch. However, this option generally doesn't provide enough benefits to make it worth the cost of setting it up.

Fix Problems in the Main Branch

If a merge operation from a development branch to the main branch causes the pipeline to fail, fix the problem in the main branch and run the fix through the pipeline. If it's a major problem, you can create a special branch dedicated solely to the problem. However, don't reuse the feature branch that was the source of the merge problem and try to fix it there. It's probable that the branch is already out of sync with the latest integration changes.

Use Release Branches as Archives

Use release branches only as archives. After the pipeline successfully verifies a change, you can store the code in a release branch before you send it to production. An archive allows you to quickly access a particular version of the code if problems occur. Make sure, however, that you always release to production the binaries that came from the main branch and that were built and validated by the pipeline.

Use the Pipeline to Perform Nonfunctional Tests

There may be many nonfunctional requirements that your application must satisfy, other than the specific behaviors detailed in the business specifications. There is an extensive list of nonfunctional requirements in the Wikipedia article *Nonfunctional requirement*. If you decide to test any of them, remember that the best way to go about it is to run the tests as either automatic or manual steps in some stage of the pipeline. For example, with the proper tools, you can run security tests as an automatic step. On the other hand, usability tests are usually performed as manual steps.

Use the Pipeline for Performance Tests

If you need to validate that your application can, for example, perform under a particular load or that it satisfies a particular capacity requirement, you can add stages to your pipeline that are devoted to performance tests.

Automate Environment Provisioning and Management

The benefits of automation is a constant theme throughout this guidance. Provisioning and managing your environments are no exceptions to this advice. If you need a single computer or an entire environment to run validations inside the pipeline, your life will be easier if the setup is done with the push of a button. Along with the usual benefits you get from automation, you will find that you can easily implement advanced test techniques such as canary releases and blue/green deployments, which are discussec later in this chapter. Virtualization, typically in the cloud, is the most convenient way to implement this automation but you can use physical machines if you have the right tools, such as the *Windows Automated Installation Kit (AIK)*.

Use Canary Releases

A canary release is when you release some version of your application to a particular group of users, before it's released to everyone else. A canary release can help you to identify problems that surface in the production environment before they affect your entire user base. There are also other benefits.

- Canary releases provide fast, pertinent feedback about your application, especially if the target users are chosen wisely.
- Canary releases can help to simplify load and capacity testing because you perform the tests against a smaller set of servers or running instances of the application.
- Canary releases make rollbacks easier because, if there are problems, you simply stop making the new version available to the target users. You also don't have to inconvenience your entire user base with the rollback.
- Once you can perform canary releases, it will be easier to add A/B testing, which is when you have two groups of target users. One group sees the new version of the software and the other group sees the old version.

To learn more about canary releases, see Chapter 10 of Jez Humble's and David Farley's book, *Continuous Delivery*.

Use A/B Testing

A/B testing consists of releasing a feature and measuring how well it performs (for example, do your users like it better than the old version, or does it improve a particular feature of the application). Depending on the results, you will either keep the feature or discard it. A/B testing is frequently done by comparing two different features that are released at the same time. Often, one is the original feature and one is the new feature. A/B testing is a powerful tool that businesses can use to get feedback about changes to the software. For more information about A/B testing, see *A/B testing* on Wikipedia. You can also refer to Chapter 10 of Jez Humble's and David Farley's book, *Continuous Delivery*.

Use Blue/Green Deployments

Blue/green deployments occur when there are two copies of the production environment, where one is traditionally named blue and the other green. Users are routed to one of the two environments. In this example, we'll say that users are routed to the green environment. The blue environment is either idle or receives new binaries from the release stage of the pipeline.

In the blue environment, you can perform any sorts of verifications you like. Once you're satisfied that the application is ready, you switch the router so that users are now sent to the blue environment, which has the latest changes. The green environment then takes on the role of the blue environment, and is either idle or verifies the code that is coming from the pipeline.

If anything goes wrong with the blue environment, you can point your users back to the green environment. Blue/green deployments mean that you can have releases with close to zero downtime, which is very useful if you are continuously (or at least, frequently) delivering new software. If you have a very large system, you can extend the blue/green model to contain multiple environments.

To learn more about blue/green deployments, see Chapter 10 of Jez Humble's and David Farley's book, *Continuous Delivery*.

Set Up Error Detection in the Production Environment

Chapter 5 discusses the importance of monitoring the application in the production environment. Catching production bugs early increases the Mean Time Between Failures (MTBF) or, to think of it in another way, lowers the defect rate.

One of the best ways to improve software quality is to make sure that the information the operations team gathers as they monitor an application is shared with the development team. The more detailed the information, the more it helps the development team to debug any problems. Good information can also help to lower the Mean Time To Recover (MTTR).

Use Telemetry and Analytics

You can lower the MTBF and the MTTR by using telemetry and analytics tools in the production environment. There are third-party tools available that look for exceptions, silent exceptions, usage trends, and patterns of failure. The tools aggregate data, analyze it and present the results. When a team examines these results, they may find previously unknown problems, or even be able to anticipate potential problems.

Purposely Cause Random Failures

It's always good to proactively look for potential points of failure in the production environment. One effective way to do this (and, in consequence, to lower the MTBF) is to deliberately cause random failures and to attack the application. The result is that vulnerabilities are discovered early. Also, the team learns how to handle similar problems so that, if they actually occur, they know how to fix them quickly.

One of the best-known toolsets for creating this type of controlled chaos is *Netflix's Simian Army*, which Netflix uses to ensure the resiliency of its own environments. Some of the disruptions it causes include:

- Randomly disabling production instances.
- Introducing artificial delays in the network.
- Shutting down nodes that don't adhere to a set of predefined best practices.
- Disposing of unused resources

Optimize Your Debugging Process

The faster you solve a problem that occurs in the production environment, the better it is for the business. The measurable effect of debugging efficiently is that your MTTR lowers. Unfortunately, problems that occur in production can be particularly difficult to reproduce and fix. The proof is that, if they were easier to detect, they would have been found sooner. There are a variety of tools available that can help you optimize your debugging process. Here are some of them.

Microsoft Test Manager

Microsoft Test Manager (MTM) allows you to file extremely detailed bug reports. For example, you can include the steps you followed that resulted in finding the bug, and event logs from the machine where the bug occurred. For an overview of how to use MTM, see *What's New in Microsoft Test Manager 2012*.

Standalone IntelliTrace

Use the standalone IntelliTrace collector to debug applications in the production (or other) environments without using Visual Studio. The collector generates a trace file that records what happened to the application. For example, it records the sequence of method calls, and the values of variables. You may be able to find the cause of the problem without rerunning the application.

Symbol Servers

A symbol server enables debuggers to automatically retrieve the correct symbol files without needing product names, releases, or build numbers. Without a symbol server, you would have to get the source code of the correct version, and search for the debugging symbols in the binaries repository (or any other repository you use). If you can't find the symbols, you would need to rebuild the application. (Note that all this work will adversely affect your MTTR.) If you do use a symbol server, you can easily retrieve the debugging information from there, on demand.

Profilers

Profilers help you to discover the source of performance related problems such as poor memory usage resource contention.

Keep the Pipeline Secure

A fully automated pipeline can be an extremely effective recipe for disaster if it's misused. For example, you can instantly disable a production server if you run an automated deployment script manually, outside of the context of the pipeline, and without knowing what the script does, or which version of the binaries or the script you're using.

Trust among team members is always the best defense. Effective teams that use continuous delivery and adopt a DevOps mindset trust each other to make changes to the environments because it's assumed they know what they're doing, they've done it many times before, and they follow the rules.

The same holds true if there are different teams for development and operations. Again, assuming that the teams use a DevOps approach, they closely collaborate with each other. In addition, they make sure that there is complete transparency about what is involved in each deployment, which changes are being made to the target environments, and any potential problems that could occur. Of course, trust is easier achieve if environments are automatically provisioned and managed, which means that a new copy of any of them is only one click away.

However, trust is generally not enough. You may have novices on your team, or there may be so many people involved that you can't know them all. Securing the pipeline is usually a necessity.

The first step to securing the pipeline is discussed in Chapter 4 of this guidance. Lock down the environments so that they can be changed only by administrators and the service accounts that run the pipeline automations. Locking down the environments prevents anyone else from purposely or accidentally logging on to the target machines and causing potentially damaging changes (perhaps by using remote scripting).

The next step is to control who can run the pipeline, or even better, who can trigger specific pipeline stages. For example, you may want particular users to have access to the UAT stage of the pipeline so that they can automatically deploy to a staging environment and perform the tests. However, it's unlikely that you want those same users to have access to the release stage, where they could potentially deploy to production. You'll need to set the permissions of the release tool (in our case, this is TFS) so that only the appropriate people can run either the pipeline, or stages of the pipeline.

The third step is to control who can modify what the pipeline does. A malicious or naive team member could introduce actions that threaten the integrity of the environments by changing what the stages, steps, or automated deployments do. Again, the available security model provided by the release tool can help to configure the appropriate permissions.

Use a Binaries Repository for Your Dependencies

Generally, you can use the TFS build drops folder as the default binaries repository but you may want to make an exception for some dependencies, such as libraries. For them, consider using the official _NuGet_ feed in Visual Studio. Using NuGet has many advantages. Team members know that they're using the correct version of these dependencies because that information is stored and managed inside the Visual Studio project and updated in version control when it changes. Furthermore, NuGet alerts you if new versions of the dependencies are available, so that you can decide if you want the updated versions.

You can get the same benefits for libraries and dependencies that are not part of the official NuGet package feed. These include, for example, third-party components and libraries that you may have purchased, for which you don't own the source code. It can also include, for example) libraries developed by your company that are common to several projects.

Simply install your own _NuGet server_ and place all these dependencies in it, and then add the feed to Visual Studio. Your feed will work the same way as the official feed. If you want, you can use a shared network folder to set up and hold the feed. Alternatively, you can set up a full featured feed that includes automatic notification of updates. If you subscribe to it, you'll be notified when a new version of the dependency is available.

To set up the repository so that it is updated automatically, create a release pipeline for the common code, have its release stage generate the NuGet packages from the validated binaries, and push the packages to the NuGet server. (Common code is code written within the organization, as opposed to third-party code, and that is used by the application in several applications. A typical example of common code is a utility class.

Use a Management Release Tool

Although not a best practice by itself, but certainly a way to help you implement some best practices is to use a release management tool whose purpose is to let you build once but deploy to multiple environments. One possibility is the _DevOps Deployment Workbench Express Edition_, which is currently available. You can read more about this tool in Appendix 1.

Another possibility is *InRelease*, which will be included with Visual Studio 2013 and is currently available as a preview. The different components that make up a pipeline and that are discussed in this guidance have counterparts in InRelease. Here are the correspondences.

- The pipeline itself is defined with a release template. Pipeline instances are called releases.
- Stages and environments are the same in InRelease as in this guidance.
- Steps are created with actions and tools artifacts. InRelease has a library with many predefined steps.
- You can define and manage additional elements, which in InRelease are referred to as components or technology types. In this guidance, these are either represented implicitly (for example, components are Visual Studio projects) or are not used in the implementation presented here.

The following screenshot shows an example of an InRelease menu that allows you to manage various components.

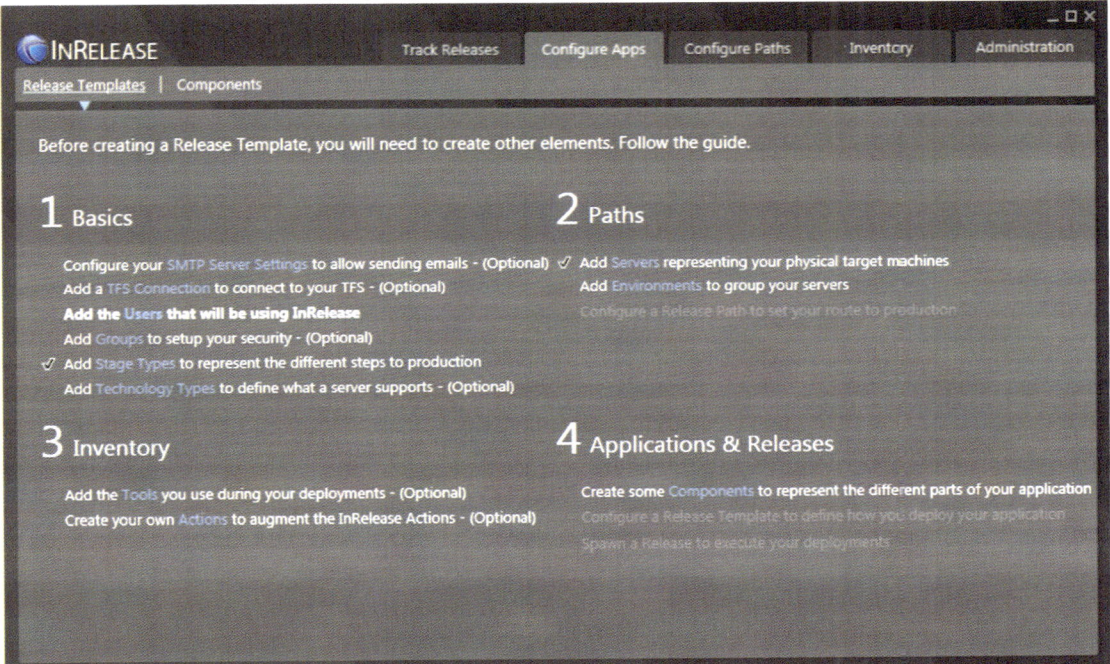

TREY RESEARCH

Now let's take a look at how Trey Research is planning to implement these patterns and practices. They still have problems they want to address in future iterations. Here are the ones that concern them the most.

Issue	Cause	Solution
They have integration problems. When they try to integrate code made by different team members or groups, there are merge conflicts. They spend too much time on integrations and it's not clear what to merge.	They use long-lived development (feature) branches.	Use feature toggles instead of feature branches.
They don't have a way to know if users like new features. This means they don't know if it's worthwhile to invest in them further or if they should discard them.	There's no mechanism in place to show new features to users while they're in the early stages of development. In other words, there's no feedback.	Use feature toggles and/or canary releases. Perform A/B tests.
They don't know if the application is vulnerable to security threats.	There are no security tests.	Introduce a security testing stage.
They don't know if the application can cope with the expected load.	There are no capacity or load tests.	Add a capacity testing stage to perform capacity and load tests.
The MTBF is too small. The system fails in production more than it should. Some of the bugs found in production should be found earlier.	The production environment isn't monitored for hard to find bugs or silent exceptions.	Integrate TFS and SCOM, so that issues detected in production by SCOM are recorded in TFS. Use analytics to gather data and identify potential issues.
The MTTR is too high. When an issue or outage occurs in production, they spend too long investigating and fixing it.	The current tools are inadequate.	Use the IntelliTrace standalone collector in the production environment to generate traces for the exceptions and to help debug them. Set up a symbol server that the commit stage updates automatically so that the debug symbols are always available to debug any build.
Team members use different or outdated versions of their own and third party libraries while developing, deploying and testing.	Code and libraries aren't managed. When a common library is updated, the people who use it aren't notified or don't have an easy way to get it.	Set up a common artifact repository for their own and third party common libraries and dependencies by using a NuGet server. For libraries they develop, update the repository automatically by using the pipeline.
It's hard to set up new environments that have all the required application and the correct configurations.	Environment provisioning is not automated.	Automate environment provisioning by using SCVMM, virtual machine templates, and Lab Management when necessary.

Of course, they know that they'll need to improve their pipeline (not once, but many times over) to finally achieve all their goals. The following diagram shows what the Trey Research pipeline will look like sometime in the future.

Future Trey Research Pipeline

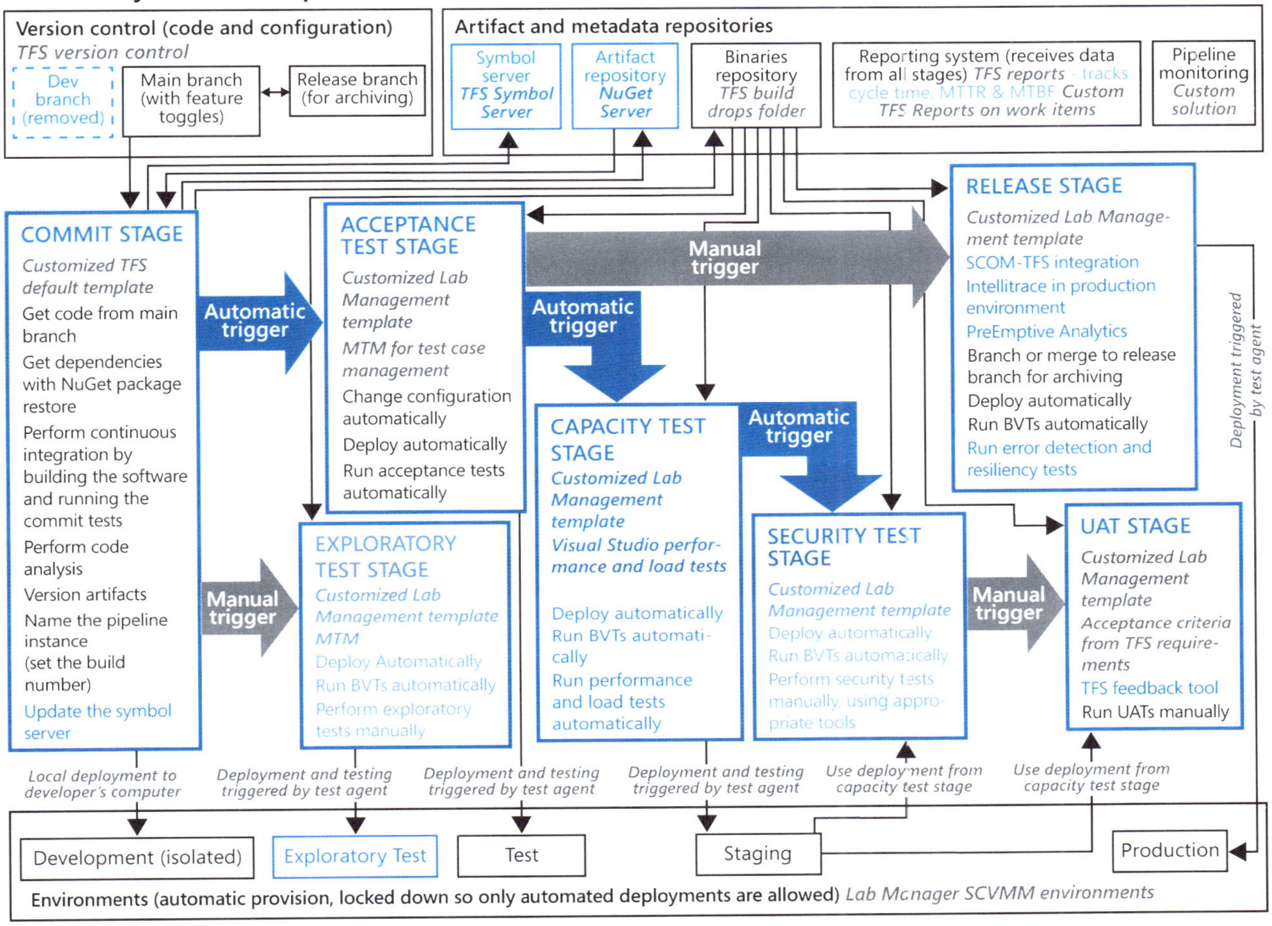

Version control (code and configuration)
TFS version control

| Dev branch (removed) | Main branch (with feature toggles) | Release branch (for archiving) |

Artifact and metadata repositories

| Symbol server **TFS Symbol Server** | Artifact repository **NuGet Server** | Binaries repository *TFS build drops folder* | Reporting system (receives data from all stages) *TFS reports* · tracks cycle time, MTTR & MTBF *Custom TFS Reports on work items* | Pipeline monitoring *Custom solution* |

COMMIT STAGE

Customized TFS default template

- Get code from main branch
- Get dependencies with NuGet package restore
- Perform continuous integration by building the software and running the commit tests
- Perform code analysis
- Version artifacts
- Name the pipeline instance (set the build number)
- Update the symbol server

Automatic trigger

ACCEPTANCE TEST STAGE

Customized Lab Management template
MTM for test case management

- Change configuration automatically
- Deploy automatically
- Run acceptance tests automatically

Automatic trigger

Manual trigger

RELEASE STAGE

Customized Lab Management template
- SCOM-TFS integration
- Intellitrace in production environment
- PreEmptive Analytics
- Branch or merge to release branch for archiving
- Deploy automatically
- Run BVTs automatically
- Run error detection and resiliency tests

Deployment triggered by test agent

Manual trigger

EXPLORATORY TEST STAGE

Customized Lab Management template MTM
- Deploy Automatically
- Run BVTs automatically
- Perform exploratory tests manually

CAPACITY TEST STAGE

Customized Lab Management template
Visual Studio performance and load tests

- Deploy automatically
- Run BVTs automatically
- Run performance and load tests automatically

Automatic trigger

SECURITY TEST STAGE

Customized Lab Management template
- Deploy automatically
- Run BVTs automatically
- Perform security tests manually, using appropriate tools

Manual trigger

UAT STAGE

Customized Lab Management template
Acceptance criteria from TFS requirements
- TFS feedback tool
- Run UATs manually

Local deployment to developer's computer | *Deployment and testing triggered by test agent* | *Deployment and testing triggered by test agent* | *Deployment and testing triggered by test agent* | *Use deployment from capacity test stage* | *Use deployment from capacity test stage*

| Development (isolated) | Exploratory Test | Test | Staging | Production |

Environments (automatic provision, locked down so only automated deployments are allowed) *Lab Manager SCVMM environments*

Jin's excited.

Friday, September 13, 2013

Everyone on the team has ideas for what we can do next. We've even created a special work item area in TFS for all the improvements! Of course, we'll have to fit them in with everything else but we can do it. The team's really a team now. This afternoon, Paulus made me a cup of coffee with his space age espresso machine. I was up until 3AM playing Halo.

Here's the current list of work items for improving the Trey Research pipeline.

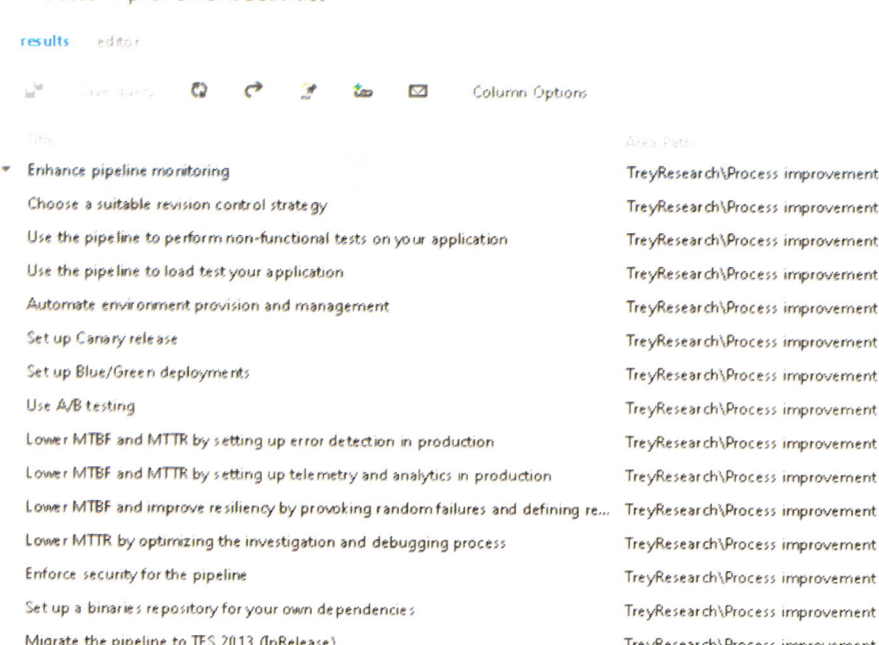

The next section discusses how Trey Research is thinking of implementing the work items.

Trey Research's Plans for Improving the Pipeline

Here's how Trey Research is thinking of implementing future improvements.

How Will Trey Research Implement Visuals

The Trey Research team is thinking about using either a web page or a Visual Studio plugin that can create these diagrams. There are several ways that they can retrieve the data.

- Team Foundation has an API that can retrieve the status of ongoing and finished builds, and also trigger new ones. For more information, see *Extending Team Foundation*.
- The *TFS OData API* exposes similar functionality.
- TFS Web Access can be extended by writing a plugin in JavaScript, although the API is still not well documented. Here's a *blog post* with a compilation of links to sites that have example plugins.

Of course, another possibility is to use a tool that already has the capability to create these visuals. One such tool is *InRelease*, which will be a part of Visual Studio 2013.

How Will Trey Research Implement a Branching Strategy

Until they become more familiar with feature toggles, Trey Research will use short-lived feature branches. They'll have frequent merges, in order to keep the integration process straightforward. In addition, they've set up a TFS alert that warns them whenever any change has been merged to the main branch. The alert lets them know that they can update whatever feature branch they're working on. They avoid cherry-picking, and the commit stage of their pipeline is configured to retrieve changes from the main branch only on every check-in. Here's an example of an alert.

How Will Trey Research Perform Nonfunctional Tests

The team's going to decide which nonfunctional tests to run on a case by case basis.

How Will Trey Research Implement Performance Tests

The team is planning to first write the performance tests. Visual Studio Ultimate has many tools that help you to write web performance and load tests. For more information, see *Testing Performance and Stress Using Visual Studio Web Performance and Load Tests*. After the tests are written, the team is planning to add at least one new stage to the pipeline that is devoted to performance tests. They will use a test rig that consists of at least one test controller to orchestrate the process, and as many test agents as necessary to generate the loads. For more information, see *Using Test Controllers and Test Agents with Load Tests*. You might also want to look at *Lab5–Adding New Stages to the Pipeline and Lab3.3– Running the Automated Tests*.

For applications with many users, you may find that it's difficult and expensive to set up a test rig that properly simulates the conditions of the production environment. A cloud-based solution, where the load is generated from agents installed in Windows Azure, may be the best way to handle this scenario. Visual Studio 2013 will provide a cloud-based load testing solution. You can get more details at the *Visual Studio ALM + Team Foundation Server blog*.

How Will Trey Research Automate Environment Provisioning and Management

Now that they're managing environments by using Lab Management, the team plans to add System Center Virtual Machine Manager (SCVMM). This will allow them to automatically manage virtual machines from within Lab Management as well as from within the stages of the pipeline. Using SCVMM will ensure that the environments are automatically provisioned when they are needed. For more information, see *Configuring Lab Management for SCVMM Environments*.

How Will Trey Research Implement Canary Releases

Performing canary releases seems quite feasible because the team has already automated deployments and tokenized the parameters. These improvements allow them to automatically configure the deployment script to suit the target environment and server that is receiving the new binaries. Raymond could configure the network infrastructure so that requests are routed to the servers dedicated to the chosen subset of users. Once they have other best practices in place, such as automated environment provisioning and feature toggles, it will be even easier to perform canary releases.

How Will Trey Research Implement A/B Testing

The team is first going to set up feature toggles and canary releases. Implementing these two best practices, along with automated deployments, will make it much easier to set up the pipeline so that it activates and de-activates specific features, or releases them only to a certain group of users.

How Will Trey Research Implement Blue/Green Deployments

As with canary releases, automated and tokenized deployments make it much easier for the team to perform blue/green deployments because they can immediately point the output of the pipeline to either environment. Once automated environment provisioning is in place, by using SCVMM, blue/green deployments will be even more feasible. With SCVMM, it will be easy to automatically set up identical environments.

How Will Trey Research Implement Error Detection

Raymond has plans to set up System Center 2012 Operations Manager (SCOM) to monitor the production environment. Once this happens, the team wants to integrate it with TFS so that issues detected by SCOM are recorded in TFS as work items. If the team updates the work items as they solve the issue, the updates are sent back to SCOM so their status is also reflected there. This keeps the development and operations teams in sync. Work items can contain useful data about the issue, such as exception details or IntelliTrace files that help the development team quickly solve the problem, keep the MTTR low and keep the operations people informed. For more information about SCOM, see _Operations Manager_. For more information about how to integrate SCOM with TFS and other tools, see _Integrating Operations with Development Process_.

How Will Trey Research Implement Telemetry and Analytics

The Trey Research team is thinking about using _PreEmptive Analytics for Team Foundation Server_. This tool ana-lyzes silent failures and automatically creates work items in TFS.

How Will Trey Research Purposely Cause Random Failures

Netflix has released the source code for the Simian Army under the Apache license, which means that the Trey Research team can modify and use it if it retains the copyright notice. The team plans to study the code and perhaps use it as the starting point for a solution that fits their situation.

How Will Trey Research Optimize Their Debugging Process

Trey Research already uses MTM to file bugs. They are learning about using IntelliTrace in the production en-vironment, symbol servers in TFS, and the Visual Studio Profiler. For more information, see _Collect IntelliTrace Data Outside Visual Studio with the Standalone Collector_. For more information about symbol servers, see _Publish Symbol Data_. For more information about profilers, see _Analyzing Application Performance by Using Profiling Tools_.

How Will Trey Research Secure Their Pipeline

In Chapter 4, the team locked down the environments so that only the appropriate service accounts and the administrator (Raymond) could log on to the machines.

For now, Raymond controls all the environments and he's still reluctant to give the rest of the team any level of access, especially to the production environment. Whether or not these rules become more relaxed remains to be seen. Many companies have policies that restrict access to the production environment to only a few people.

Raymond has enforced his control of the environments by denying permission to manually queue build definitions in TFS. This restriction won't impact the rate that at which the team delivers releases because, in TFS, each check-in triggers a new pipeline instance that begins with the commit stage build definition. The pipeline instance is created independently of whether the person doing the check-in has permission to manually queue builds or not.

The drawback is that, if each automatically triggered stage succeeds, the pipeline advances until it reaches the first manually triggered stage. At that point, the team member or person who performs the UATs must ask Raymond to trigger it for them. If the release schedule becomes very busy Raymond may become a bottleneck, which will be reflected in the Kanban board. He may have to adopt a different strategy, such as allowing another team member to also trigger the stages.

Even if everyone on the team could queue builds, they still need to restrict access to the pipeline stages. For example, they won't want their UAT testers to have access to any stage but the UAT stage. Fortunately, in TFS, it's possible to set permissions at the build definition level, so you can let users trigger a specific stage while stopping them from triggering any others. The following screenshot shows how to use Team Explorer to access the build definition permissions by selecting **Security**.

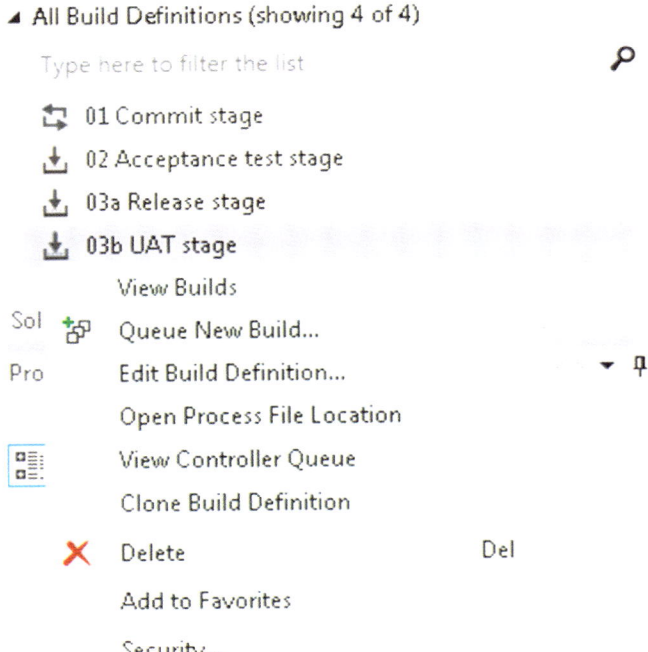

The permissions that determine whether someone can or cannot change any component of the pipeline can be so granular because everything is under version control. Permissions in version control can be set at the item level, where an item can be the customized build process templates that make up the stages, all the scripts, and the code used by the steps.

Will Trey Research Implement a Binaries Repository

The team plans to investigate how to create a script that generates the NuGet packages and pushes them to the NuGet server. They think they'll use the *nuget.exe command line tool*, and the pack and push commands.

THE CONCLUSION

Here are Jin's final thoughts, at least for now.

MORE INFORMATION

There are a number of resources listed in text throughout the book. These resources will provide additional background, bring you up to speed on various technologies, and so forth. For your convenience, there is a bibliography online that contains all the links so that these resources are just a click away. You can find the bibliography at: *http://msdn.microsoft.com/library/dn449954.aspx*.

If you're not familiar with branching strategies, and specifically with feature branches, see the Rangers' *Visual Studio Team Foundation Server Branching and Merging Guide*. However, if you're using continuous delivery, apply its advice after you've made sure you're following the best practices in this guidance.

For information on how to use Git with TFS, see Brian Harry's blog at *http://blogs.msdn.com/b/bharry/archive/2013/06/19/enterprise-grade-git.aspx*.

You can read more about branching in the context of continuous delivery in Jez Humble's article DVCS, continuous integration, and feature branches at *http://continuousdelivery.com/2011/07/on-dvcs-continuous-integration-and-feature-branches/*.

A good place to start learning about feature toggles is Martin Fowler's blog at *http://martinfowler.com/bliki/FeatureToggle.html.*

There is an extensive list of nonfunctional requirements in the Wikipedia article Nonfunctional requirement at *https://en.wikipedia.org/wiki/Non-functional_requirement*.

Virtualization, typically in the cloud, is the most convenient way to implement this automation but you can use physical machines if you have the right tools, such as the Windows Automated Installation Kit (AIK) available at _http://www.microsoft.com/en-us/download/details.aspx?id=5753_.

More information about A/B testing is available A/B testing on Wikipedia at _http://en.wikipedia.org/wiki/A/B_testing_.

One of the best-known toolsets for creating this type of controlled chaos is Netflix's Simian Army at _http://techblog.netflix.com/2011/07/netflix-simian-army.html_, which Netflix uses to ensure the resiliency of its own environments.

For an overview of how to use MTM, see What's New in Microsoft Test Manager 2012 at _http://msdn.microsoft.com/magazine/jj618301.aspx_.

Generally, you can use the TFS build drops folder as the default binaries repository but you may want to make an exception for some dependencies, such as libraries. For them, consider using the official NuGet feed in Visual Studio at _www.nuget.org_.

To install your own NuGet server and place all these dependencies in it, and then add the feed to Visual Studio available at _http://www.nuget.org/packages/NuGet.Server/_.

Use a release management tool whose purpose is to let you build once but deploy to multiple environments. One possibility is the DevOps Deployment Workbench Express Edition available at _https://vsardevops.codeplex.com/_.

For information about InRelease, which will be included with Visual Studio 2013 and is currently available as a preview see _http://www.incyclesoftware.com/inrelease/inrelease-2013-preview/_.

For more information on using Team Foundation to retrieve the status of ongoing and finished builds, and also trigger new ones , see Extending Team Foundation at _http://msdn.microsoft.com/library/bb130146(v=vs.110).aspx_.

For more information about the TFS OData API, see _https://tfsodata.visualstudio.com/_.

TFS Web Access can be extended by writing a plugin in JavaScript, although the API is still not well documented. Here's a blog post with a compilation of links to sites that have example plugins _http://bzbetty.blogspot.com.es/2012/09/tfs-2012-web-access-plugins.html_.

For more information about the tools available in Visual Studio Ultimate that help you to write web performance and load tests, see Testing Performance and Stress Using Visual Studio Web Performance and Load Tests at _http://msdn.microsoft.com/library/vstudio/dd293540.aspx_.

For more information on using a test rig that consists of at least one test controller to orchestrate the process, see Using Test Controllers and Test Agents with Load Tests at _http://msdn.microsoft.com/library/vstudio/ee390841.aspx_.

For more information on using SCVMM to ensure that the environments are automatically provisioned when they are needed, see Configuring Lab Management for SCVMM Environments at _http://msdn.microsoft.com/library/vstudio/dd380687.aspx_.

For more information about SCOM, see Operations Manager at _http://technet.microsoft.com/library/hh205987.aspx_.

For more information about how to integrate SCOM with TFS and other tools, see Integrating Operations with Development Process at _http://technet.microsoft.com/library/jj614609.aspx_.

For more information about using PreEmptive Analytics for Team Foundation Server to analyze silent failures and automatically create work items in TFS, see _http://www.preemptive.com/products/patfs/overview_.

For more information on using IntelliTrace in the production environment, see Collect IntelliTrace Data Outside Visual Studio with the Standalone Collector at _http://msdn.microsoft.com/library/vstudio/hh398365.aspx_.

For more information about symbol servers, see Publish Symbol Data at *http://msdn.microsoft.com/library/hh190722.aspx*.

For more information about profilers, see Analyzing Application Performance by Using Profiling Tools at *http://msdn.microsoft.com/library/z9z62c29.aspx*.

For information about using the nuget.exe command line tool, and the pack and push commands see *http://docs.nuget.org/docs/reference/command-line-reference*.

The hands-on labs that accompany this guidance are available on the Microsoft Download Center at *http://go.microsoft.com/fwlink/p/?LinkID=317536*.

Appendix 1

DevOps Deployment Workbench Express Edition

This appendix discusses the DevOps Deployment Workbench Express Edition (Express Edition). The tool allows you to use a single build and deploy it to many different environments. In other words, in keeping with the best practices for continuous delivery, the workbench allows you to build once and deploy anywhere. The Express Edition accomplishes this by making it easy both to create deployment processes, called orchestrations, and to monitor and manage your deployments. The tool has two components. The DevOps Workbench provides XAML workflow items that you use to create the orchestrations. Similar to the workflow designer in Microsoft Visual Studio, you select items from a toolbox and drag them to a design surface. When deployment processes exist as code rather than as documents, they become versionable, repeatable, and predictable.

The second component, Deployment Explorer, is similar to Build Explorer and is available from the Team Explorer tab in Visual Studio. With Deployment Explorer you can manage deployments and review the results of completed deployments.

Here are the major subjects covered in this appendix.

- **Installing the Express Edition.** This section includes the prerequisites and the installation procedure.
- **Constructing an Orchestration.** This section first shows you how to use the workbench to create a simple orchestration and then shows you how to add complexity to the orchestration.
- **Deploying a Build.** This section shows you how to deploy a build by using the workbench, the command line, or Deployment Explorer.
- **Using DevOps Workbench to Build Once, Deploy Anywhere.** This section shows you how the workbench can help you implement a continuous delivery release process.

INSTALLING THE EXPRESS EDITION

This section includes the software and hardware prerequisites for installing the Express Edition as well as the instructions for installing it on your local machine.

Software Prerequisites

This table lists the software prerequisites for the two components that make up the Express Edition, the DevOps Workbench and the Deployment Explorer.

Component	Software
DevOps Workbench	Windows Server 2008 R2, Windows Enterprise R2 64 bit (IIS installed), Windows Server 2012, Windows 7, Windows 8 or Windows 8.1
	Team Foundation Server 2010 or 2012
	Build controller
	Build agent
Deployment Explorer	Microsoft Visual Studio 2010 or 2012, Professional, Premium, Ultimate, or Test Professional

Hardware Prerequisites

You install the Express Edition on your local machine. You will need another machine to act as the target for the deployment.

Installing the Express Edition

In this section you download and install the Express Edition.

1. Download the Express Edition zip file from the _Visual Studio ALM Rangers CodePlex site_ to a location on your local machine.
2. Open the zip file. You should see the following files.
   - DevOps Deployment Workbench Express MSI
     - Workbench
     - Deployment Explorer
     - WFExecute.exe
   - HOL folder
   - Guidance folder
3. Select the DevOpsDeploymentExpress.msi file. Follow the steps in the Install Wizard.

Creating Repeatable Builds

A build should be a repeatable, reliable process. It produces a set of binaries that have been validated by a series of tests, and that is documented by a build record in Team Foundation Server. The Express Edition requires that there is a single build that has been validated to run correctly, and that is in a secure drop location. The tool retrieves the build and then deploys it, using the steps in the orchestration, to the target environments.

To learn more about how to create builds, see the information in the guidance folder that is included in the DevOps Workbench download, or see the _Team Foundation Build Customization Guide_ for more information on creating reusable build definitions.

CONSTRUCTING AN ORCHESTRATION

In this section you familiarize yourself with the DevOps Workbench user interface (UI) and learn how to create an orchestration.

Understanding the DevOps Workbench UI

The DevOps Workbench UI has three areas (numbers refer to the numbers shown in the screenshot). The **Deployment Toolbox** (1) contains all the items that you need to construct an orchestration. The central section (2) is the design surface where you drag deployment templates and components to construct the orchestration. The **Properties** area (3) allows you to view and modify the properties of the items that comprise the orchestration. The following screenshot shows the DevOps Workbench UI.

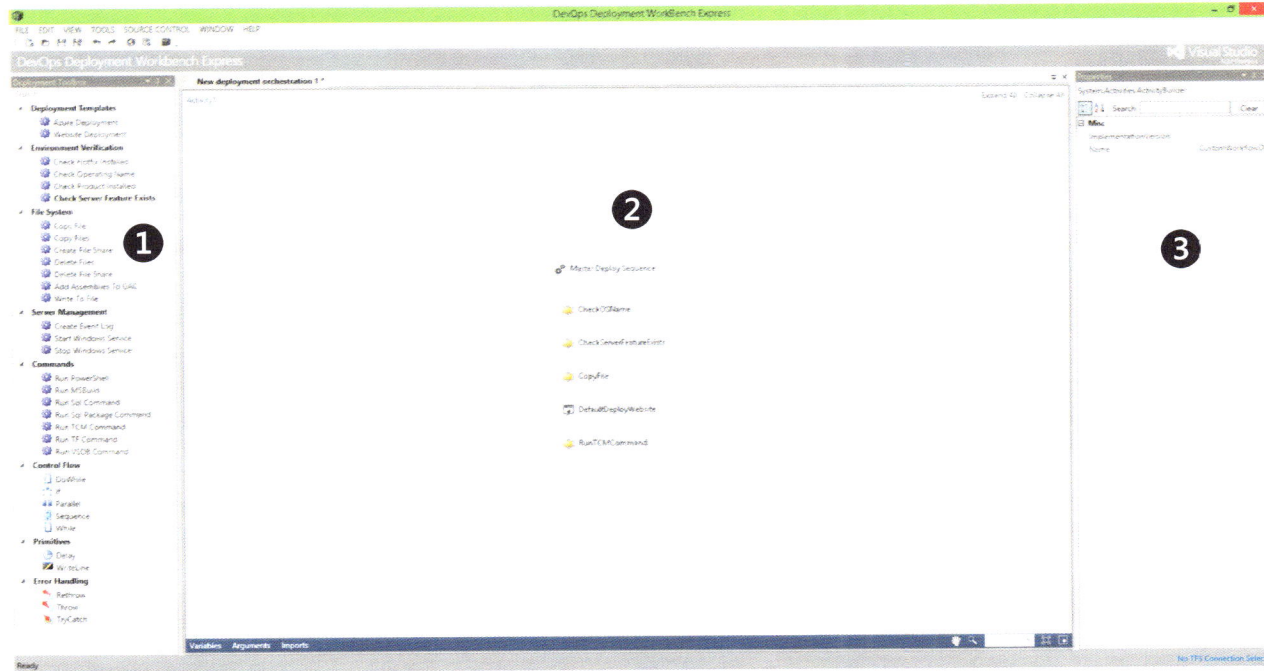

Creating an Orchestration

In this section you learn how to create an orchestration. You'll need one orchestration per server that you'll deploy to by using the Express Edition.

1. Open Express Edition. Click **File**. Click **New Orchestration** (Ctrl + N).

2. From the **Toolbox**, from the **Environment Verification** section, select the **CheckOSName** activity and place it in the **MasterDeploy** sequence, which is on the design surface by default. The **CheckOSName** activity verifies that the target machine exists and ensures that the appropriate operating system is installed.

3. Select the **DefaultDeployWebsite** activity and place it after the **CheckOSName** activity.

4. Close and save the orchestration XAML File.

The following screenshot shows the complete orchestration.

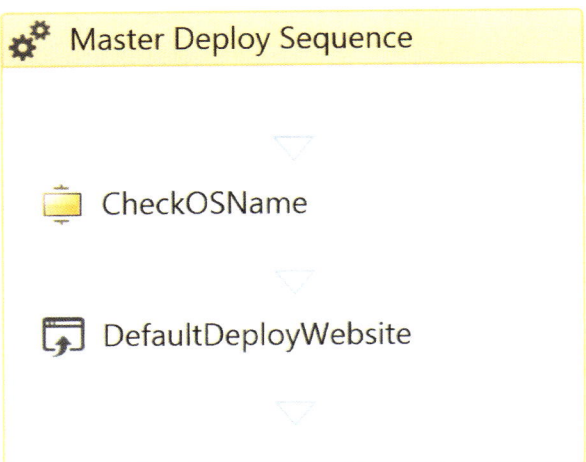

Adding Items to an Orchestration

You can add as many items as you need to your orchestration in order to configure the servers you're deploying to, and to install the prerequisites that your application depends upon.

The following orchestration expands on the previous example. It deploys a website, Microsoft SQL Server, and some test cases that will run on the target machine after the deployment is complete.

Within the **Master Deploy Sequence**, the **Website** item includes an activity that first checks that the operating system of the target machine is the correct version. The orchestration then deploys the website. Next, the **SQL Server** item first verifies that the operating system is the correct version and then installs and configures SQL Server on the target machine. Finally, some validation tests run to ensure that the deployment succeeded. The following screenshot shows the completed orchestration.

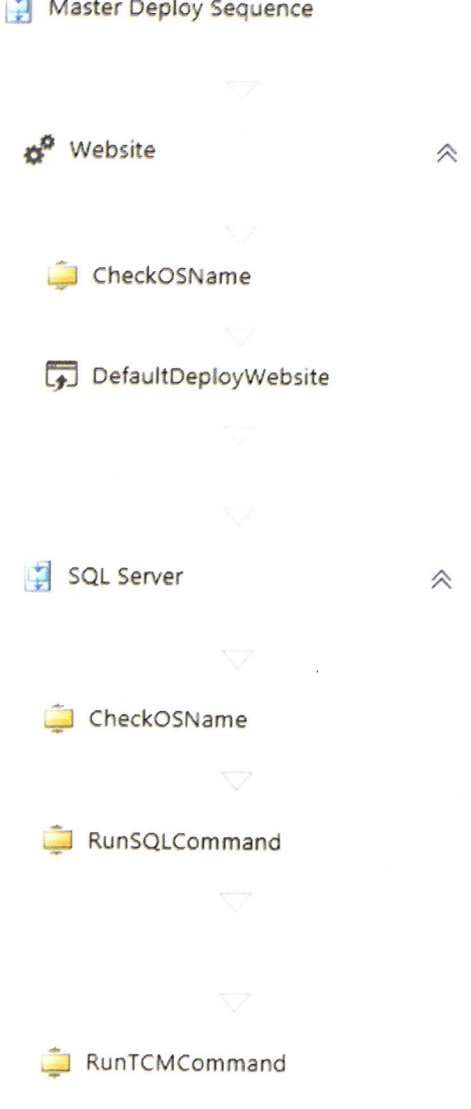

Properties

Each of the items that comprise an orchestration has properties that need to be defined. Some of them are required, while others are optional. You can hover over an element to see its tooltip, or you can read the _ALM Rangers DevOps Quick Reference Guide_ to get more information and the syntax for each item.

As an example, select **CheckOSName**. You can see its properties in the **Properties** area, as shown in the following screenshot.

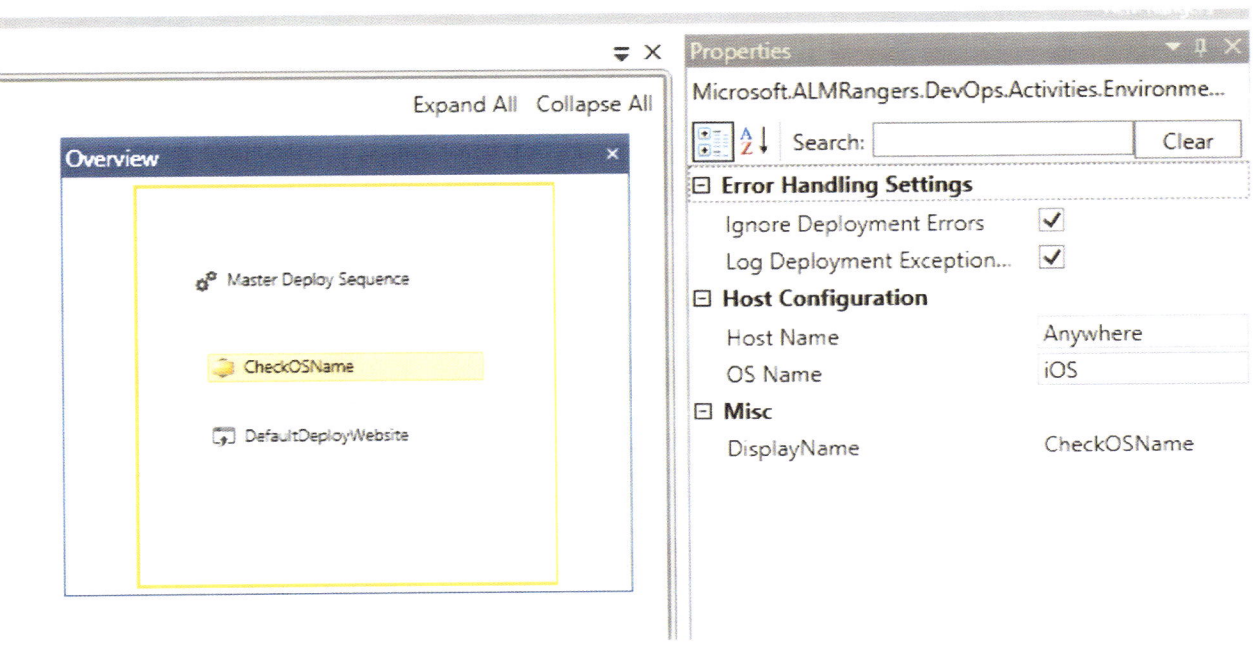

One of the properties is **Ignore Deployment Errors**. Clear this option if you do not want to ignore the errors. If you want to log any deployment exceptions, select **Log Deployment Exceptions**. After updating the properties, save the orchestration XAML file.

ENABLING THE CREDENTIAL SECURITY SUPPORT PROVIDER

Because the Express Edition uses remote PowerShell, it can double hop from the local machine to the target machine. This means that the credential security support provider (CredSSP) must be **ON**, on both the server and the client.

1. Open a Windows PowerShell prompt. To enable the client-side CredSSP, enter the following command.

```PowerShell
Enable-WSManCredSSP -Role client -DelegateComputer *
```

2. To enable the server-side CredSSP, enter the following command.

```PowerShell
Enable-WSManCredSSP -Role server
```

For more information, see _Enable—WSManCredSSP_.

DEPLOYING A BUILD

There are three ways to deploy a build. One is to use the DevOps Workbench, another is to use the command line, and the third is to use Deployment Explorer.

Deploying by Using the DevOps Workbench

This section shows you how to deploy by using the DevOps workbench.

1. Make sure you're connected to TFS. Click **Source Control**. Click **Change Connection**.

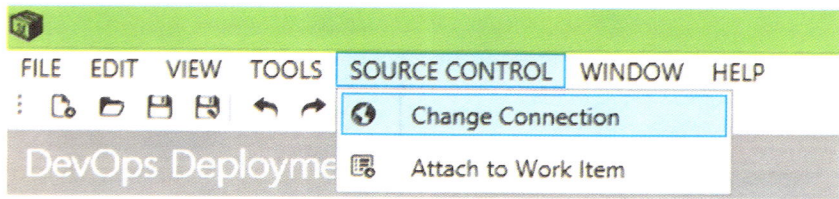

2. You should see a **Connected to** message in the lower right hand corner of the screen. The following screenshot shows an example.

3. From the **Tools** menu, click **Target Servers** (Ctrl + Shift + S). The **View Target Servers** dialog box appears. Click **Add**. Enter the server name, the IP Address, the user name and the password.

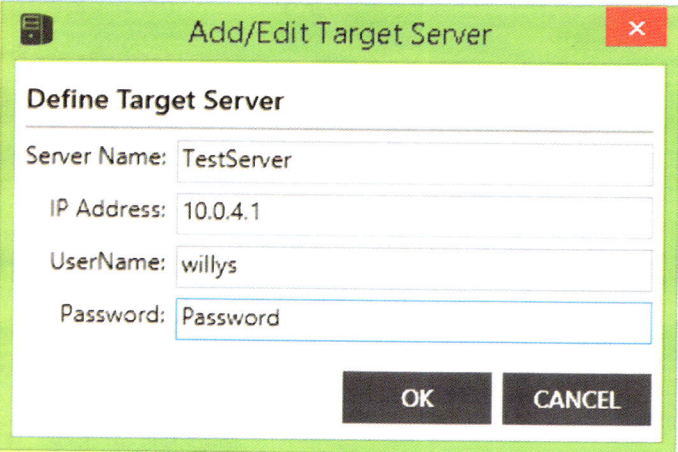

4. Click **OK**. Click **Refresh**. The **View Target Servers** dialog box reappears. The new server is included in the target server list. The following screenshot shows an example.

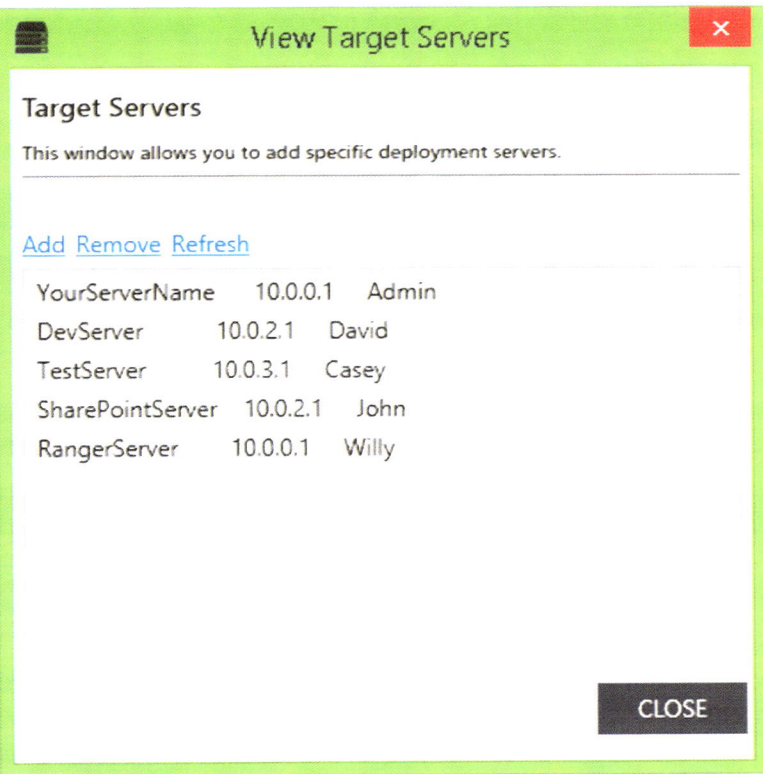

5. Click **Close**.
6. From the **Tools** menu, select **Deploy to Server**. The following screenshot shows an example.

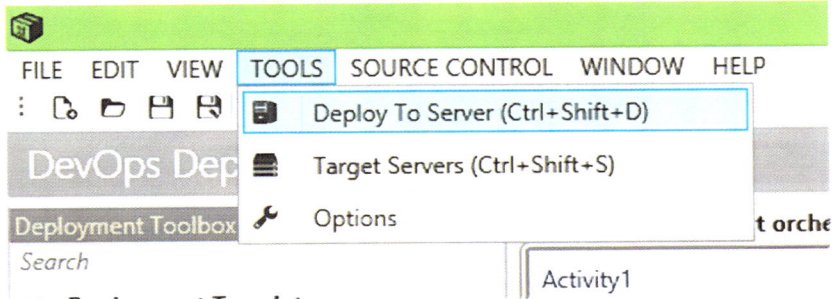

7. The Deploy **Package To Server** dialog box appears. In the **Package Name** text box, enter a package name.
8. In the **Package Xaml** text box, enter the name of the orchestration XAML file that you saved in "Creating an Orchestration."
9. In the **Target Server** text box, enter the name of the target server that you connected to in step 1. If you want to use the IP address of that server instead, select that option and enter the IP address in the **Target Server** text box.

10. In the **Target Build** text box, browse to the location of the build in TFS that you want to deploy. The following screenshot shows a completed example.

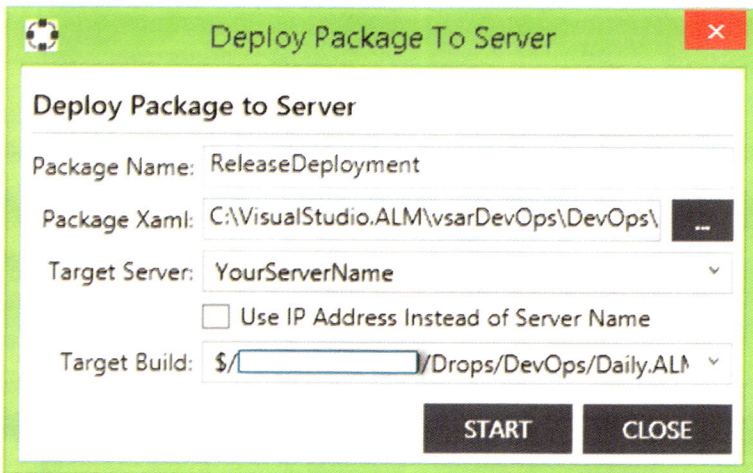

11. Click **Start**. The DevOps Workbench executes the deployment by using remote PowerShell to access the target machine, where it will run each of the orchestration steps that you created. The following screenshot shows an example of a deployment that is in progress.

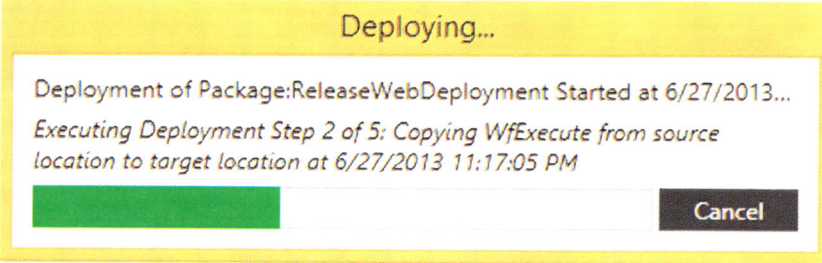

12. When the deployment completes you will get a message saying that the deployment was successful (or not), and a log file that contains your results appears.

Deploying by Using the Command Line

If you don't want to rely on the DevOps Workbench UI to deploy your application, then you can use the command line and the WFExecute.exe file, which is in the installation directory, to execute the XAML file that you created.

To use **WFExecute**, you need to know the syntax.

• All the command line tasks are parsed and have the following key/value pair for the designated parameter name: <<**Parametername**>>:<<**Propertyname**>>=<<**Propertyvalue**>> (no spaces).

• Currently there are only two designators possible for **Parametername**. The designator can be either '/p' or '/source'. For the source **Propertyname** there are only two valid values. One is **file** and the other is **assembly**.

As an example, assume that the XAML deployment file is saved as "C:\XAMLFiles\MasterDeployment.xaml". Here is the correct command.

```
CMWFExecute.exe "/source:file=C:\XAMLFiles\MasterDeployment.xaml"
```

> *Note: You must use the fully qualified path of the XAML file that you saved.*

If you want to deploy a particular assembly, you can use **WFExecute** to deploy just the assembly and activity that you want. Here is the syntax for the command line.

```
WFExecute.exe"/source:assembly=Fullyqualifiedpathtoassembly" "/p:activity=nameofactivityinassembly"
```

> *Note: You must specify all the arguments for the assembly. The format is "/p:[argumentname]=value". If you need to find out what the arguments are you can either look in the guidance at the ALM Rangers DevOps Tooling and Guidance or look at the assembly's properties by using Deployment Explorer.*

Deploying by Using Deployment Explorer

Using Deployment Explorer to deploy a build is similar to using the Team Explorer **Builds** menu. To access Deployment Explorer, open Visual Studio. Select **Team Explorer**. You will see that there is a new selection available named **Deployments**. Similar to the standard **Builds** selection, **Deployments** allows you to manage and organize your deployments.

The following screenshot shows the Team Explorer window when Deployment Explorer is installed.

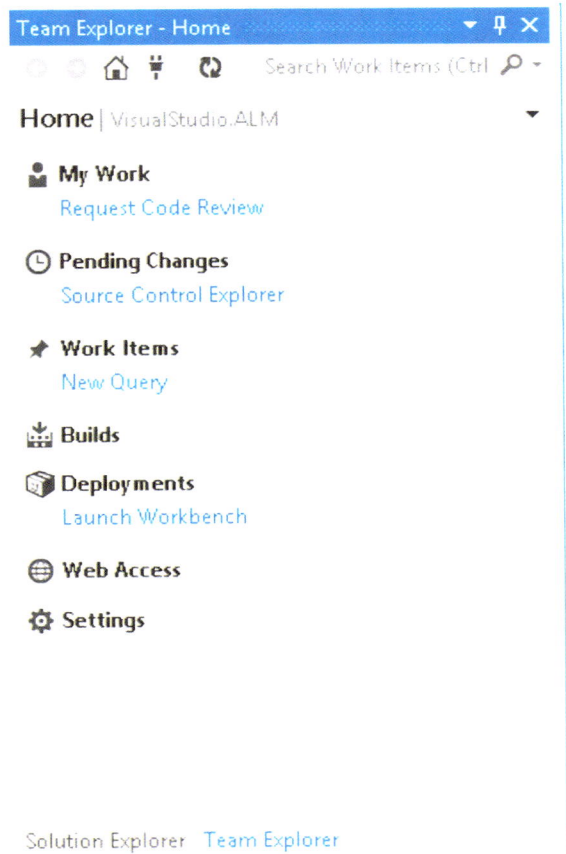

To use Deployment Explorer, simply click **Deployments**. You will see your recent deployments. The following screenshot shows an example.

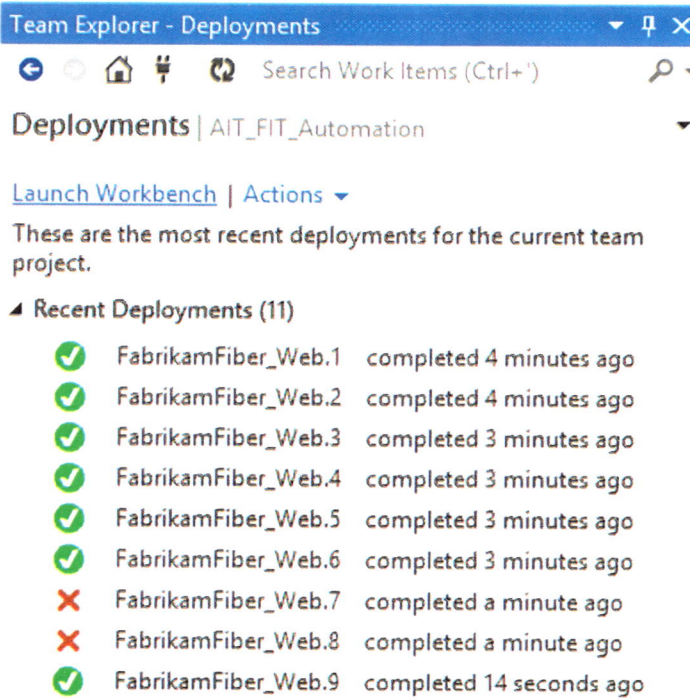

From Deployment Explorer, you can select a particular deployment to see details about it and to see the deployment log files. If you need to return to the DevOps Workbench, perhaps to fix some errors or to improve the orchestration, click **Launch Workbench**, under **Deployments**.

USING THE DEVOPS WORKBENCH TO BUILD ONCE, DEPLOY ANYWHERE

The guidance that accompanies Express Edition will help you to understand how to incorporate DevOps principles into a continuous delivery release process. You can download the *ALM Ranger DevOps Deployment poster*, which illustrates the DevOps approach.

Creating Suitable Builds

In order to take full advantage of the Express Edition, you should create builds that conform to best practices for continuous delivery. The goal is to build only once, and to deploy the same binaries to all your environments. A single build ensures that the build you tested in all the stages of the release pipeline is the build that is deployed to your customers. This is called *build integrity*. As has been shown in the rest of this guidance, separate builds for each environment can cause many problems.

One way to ensure build integrity is to lock the build share down by making it read only and to give write permission only to the build account. This practice guarantees that all changes are in source control, and that no one can circumvent the rules by adding unauthorized changes to the code or to the configuration.

You might consider using gated check-ins, which ensure that only code that meets the automated verification steps included in the build process is committed. Another approach, which is used in the HOLs for this guidance, is to use continuous integration as the trigger. What is important is that you receive immediate feedback on whether the build was successful or not.

Creating build definitions that conform to patterns and best practices and having a regular build schedule gives you confidence that the builds will either compile or be straightforward to fix, that successful builds will be in a secure drop location, that the builds are clearly identified and can be traced back to the source, and that other stages will always use the authorized build and not create builds of their own.

Using the Express Edition to Manage Deployments

As you've seen, the Express Edition can help you to conform to the best practices for continuous delivery. For example, using the tool ensures that you use the same binary to deploy to multiple environments because you can only retrieve the build from the drop location. It also creates standard ways to deploy to different environments, such as a website, SQL Server or Windows Azure. Again, as you've seen in the rest of this guidance, standardized deployments that ensure that the environments are automatically configured correctly get rid of many of the problems that make releases to production so difficult.

MORE INFORMATION

There are a number of resources listed in text throughout the book. These resources will provide additional background, bring you up to speed on various technologies, and so forth. For your convenience, there is a bibliography online that contains all the links so that these resources are just a click away. You can find the bibliography at: *http://msdn.microsoft.com/library/dn449954.aspx*.

ALM Rangers DevOps Tooling and Guidance at *https://vsardevops.codeplex.com/*.

Team Foundation Build Customization Guide at *http://vsarbuildguide.codeplex.com/*.

Visual Studio ALM Rangers Solutions Catalogue at *http://aka.ms/vsarsolutions*.

Visual Studio ALM Rangers Visual Studio Lab Management Guide at *http://vsarlabman.codeplex.com/*.

Installing Team Foundation Server and Visual Studio ALM at *http://msdn.microsoft.com/library/dd631902.aspx*.

The hands-on labs that accompany this guidance are available on the Microsoft Download Center at *http://go.microsoft.com/fwlink/p/?LinkID=317536*.

Glossary

A/B testing – a method of testing that uses two versions of the entity to be tested, such as a web page, to determine which of the two generates more interest. One version of the entity is the control and the other is the treatment. The two are identical except for a single variation. The control is the original version and the treatment has the modification. To use a web page as an example, the control is shown to some group of users and the treatment to another group of the same size. The two pages are monitored to see which generates *the most clicks*.

Agile – a group of software development methods based on iterative and incremental development, where requirements and solutions evolve through collaboration. It promotes adaptive planning, evolutionary development and delivery, a time-boxed iterative approach, and encourages rapid and flexible response to change. The *Agile Manifesto* introduced the term in 2001.

Application lifestyle management (ALM) – a continuous process of managing the life of an application through governance, development, and maintenance. ALM is the marriage of business management to software engineering made possible by tools that facilitate and integrate requirements management, architecture, coding, testing, tracking, and release management. (Wikipedia)

Branch – a copy of a set of files in a different part of the repository that allows two or more teams of people to work on the same part of a project in parallel.

Build – preparing all the artifacts that will be run by the steps included in a stage. In other words, a build does more than compile the source code. It can also include activities such as copying files, references, and dependencies, including supporting files such as images and software packages, and signing assemblies.

Build agent – in terms of TFS, the build agent does the processor-intensive and disk-intensive work. This work includes getting files from and checking files into version control, provisioning the workspace, compiling the code, and running tests. Each build agent is dedicated to and controlled by a single build controller.

Build automation – the practice of triggering a build programmatically on the occurrence of some event, such as a check-in to the version control system. A common way to automate a build is to use scripting. A single command runs the entire build. An automated build compiles source code, packages the compiled code, and can also run tests, deploy the software to different environments, and create documentation.

Build controller – in terms of TFS, a build controller accepts build requests from any team project in a specified team project collection. Each build controller is dedicated to a single team project collection. Each build controller pools and manages the services of one or more build agents. It distributes the processor-intensive work (such as compiling code or running tests) to the build agents within its pool. The build controller processes the workflow and typically performs mostly lightweight work such as determining the name of the build, creating the label in version control, logging notes, and reporting status from the build.

Build definition – in terms of TFS, a build definition contains instructions about which code projects to compile, which additional operations to perform, and how to perform them.

Build machine – in terms of Team Foundation Server, a build machine is a computer on which you have installed and configured Team Foundation Build Service. The machine can be a physical machine or a virtual machine.

Build script – the commands used to create an automated build. Shell scripts, batch files, and PowerShell scripts are examples of ways to create a build script. There are also specialized build automation tools available such as MSBuild, Make, Ant, and Maven.

Build verification tests (BVT) – a group of tests used to determine the health of a build at a high level. Typically, these tests exercise the core functionality to help team members determine whether further testing is worthwhile. They are run after the daily build to verify that compilation of source code has been built successfully and is ready for further testing. Also known as *smoke tests*.

Change – see *Version*.

Commit stage – the stage where the binaries and artifacts are built.

Commit test – a test run during the *commit stage* of the pipeline.

Continuous delivery – a pattern to improve the process of delivering software. The deployment pipeline is seen as a series of validations through which the software much pass before it is considered "done." With CD, the implication is that software is always ready to be released. An actual release occurs in accordance with business needs, but any commit can be released to customers at any time.

Continuous integration – a pattern of merging development branches into the main branch. CI automatically occurs on a build server every time there is a commit to the revision control system, and automated unit and integration tests run after the code compiles. The pattern can be extended to include the automated run of other tests, such as functional tests.

Cycle time – a metric that shows the time from when you implement a change to when it is ready to release.

Defect rate – the inverse of MTBF. The number of defects found per unit of time.

DevOps – an approach to software development that advocates a collaborative working relationship between development and information technology (IT) operations. Its goals are to achieve high deployment rates and to increase the reliability and stability of the production environment. It is closely tied to the idea of continuous delivery.

Kanban – a method for developing software products and processes with an emphasis on just-in-time delivery while not overloading the software team. In this approach, the process, from definition of a task to its delivery to the customer, is displayed for on a *kanban board*. The board allows participants to see and pull work from a queue.

Kanban board – a visual tool that represents the steps that make up the development process, from the formulation of the idea to the delivery of the product or service to users. A common way to create a kanban board is to use a wall or board and sticky notes to represent each step. Kanban boards may be derived from the *value stream*.

Lead time – the time it takes from when you first consider making a change to when you can release it. Unlike cycle time, lead time includes activities not associated with development, such as evaluating whether the feature is worth implementing.

Lean software development – a subculture within the *Agile* community. It adapts principles used by Toyota for its manufacturing processes to software development.

Mean Time Between Failure (MTBF) – average length of time the application runs before failing.

Mean Time to Recovery (MTTR) – average length of time needed to repair and restore service after a failure.

Merge – occurs when you take code from two branches and combine them into a single codebase.

Orchestration – the automated arrangement, coordination, and management of the release pipeline.

Pipeline – an implementation of the company's software release process. In continuous delivery, any instance of a pipeline supports the validation of a single version or change.

Pipeline stage – represents a set of related activities or steps that are taken during the software development process. Typical examples of pipeline stages are the commit stage, the automated acceptance test stage and the release stage.

Pipeline step – an activity that takes place within the context of a *pipeline stage*. Examples of pipeline stages are building the code and running tests.

Reverse integration – occurs when you merge code from a child branch to the parent branch.

Scrum – an iterative and incremental agile software development framework for managing software projects and product or application development. Its focus is on "a flexible, holistic product development strategy where a development team works as a unit to reach a common goal" as opposed to a traditional, sequential approach.

Smoke test – see *Build Verification Test*.

System Center Virtual Machine Manager (SCVMM) – the portion of Microsoft System Center used to orchestrate virtual machine operations such as deployment, provisioning, snapshots and state management, across one or more physical Hyper-V host machines.

Test-driven development (TDD) – a pattern that relies on very short development cycles: a developer writes an (initially failing) automated test case that defines a desired improvement or new function, then produces the minimum amount of code to pass that test, and finally refactors the code to remove any duplication and improve the design. After every refactoring, the developer runs the tests again.

Unit test – tests a particular piece of code in isolation. Unit tests should not, for example, call a database or talk to external systems. This isolation allows unit tests to run quickly, so that you get immediate feedback on the build.

User acceptance testing (UAT) – tests that are often performed as the last phase of testing. During UAT, actual software users test the software to make sure it can handle required tasks in real world scenarios, according to specifications.

User story – describes a feature, service, or product in a way that is comprehensible to a nontechnical stakeholder. It contains a written description or short title, conversations about the story that flesh out the details of the story, and acceptance tests that document details that can be used to determine when a story is complete.

Value stream – all the steps required to take a product or service from its initial state to the customer. It includes all the people, processes, times, information, and materials that are included in the end-to-end process.

Value stream mapping – a flow diagram that shows every step in the value stream.

Version – a piece of software in a unique state. In this book, a change and a version are interchangeable terms. A change is some modification in the code or supporting artifacts that is checked in to version control. Because every change is a candidate to be released into production, it is also a version.

Version number – a unique number assigned to a version of software.

Version control system – a mechanism for maintaining multiple versions of files so that previous versions of modified files are still available. A principle of the DevOps methodology is that everything should be in the same version control system. This includes the standard files, such as source code, but should also include every artifact required to build the software. For example, build and deployment scripts, configuration files, libraries, and database scripts should all be in the version control system.

Visual Studio Lab Management – a TFS capability that allows you to orchestrate physical and virtual test labs, self-provision environments, automate build-deploy-test workflows, and encourage collaboration between testers and developers.

Visual Studio Team Foundation Server 2012 (TFS) – the collaboration platform at the core of Microsoft's ALM solution. TFS supports agile development practices, multiple IDEs and platforms locally or in the cloud. It provides tools to manage software development projects throughout the IT lifecycle. These tools include source control, data collection, reporting, and project tracking.

Wait time – the time in the development cycle when there are delays and no value is added.

Windows Installer (MSI) – a software component for Microsoft Windows computers that is used to install, maintain, and remove software.

Windows Installer XML (WiX) – a toolset that builds   packages from an XML document.

XAP file – used to install apps on the Windows Phone 8.

Index

Made in the USA
San Bernardino, CA
04 January 2014